Lecture Notes in Computer Science 7309

Commenced Publication in 1973
Founding and Former Series Editors:
Gerhard Goos, Juris Hartmanis, and Jan van Leeuwen

T0215922

Massimo Ferri   Patrizio Frosini
Claudia Landi   Andrea Cerri
Barbara Di Fabio (Eds.)

Computational Topology in Image Context

4th International Workshop, CTIC 2012
Bertinoro, Italy, May 28-30, 2012
Proceedings

 Springer

Volume Editors

Massimo Ferri
Patrizio Frosini
Università di Bologna
Dipartimento di Matematica
Piazza di Porta S. Donato, 5, 40126 Bologna, Italy
E-mail: {massimo.ferri,patrizio.frosini}@unibo.it

Claudia Landi
Università di Modena e Reggio Emilia
Dipartimento di Scienze e Metodi dell'Ingegneria
Via Amendola 2, Pad. Morselli, 42122 Reggio Emilia, Italy
E-mail: clandi@unimore.it

Andrea Cerri
Barbara Di Fabio
Università di Bologna
ARCES
Via Toffano 2/2, 40125 Bologna, Italy
E-mail: {andrea.cerri2,barbara.difabio}@unibo.it

ISSN 0302-9743   e-ISSN 1611-3349
ISBN 978-3-642-30237-4   e-ISBN 978-3-642-30238-1
DOI 10.1007/978-3-642-30238-1
Springer Heidelberg Dordrecht London New York

Library of Congress Control Number: 2012937078

CR Subject Classification (1998): I.4, I.3.5, G.2, F.2, I.5-6

LNCS Sublibrary: SL 1 – Theoretical Computer Science and General Issues

Typesetting: Camera-ready by author, data conversion by Scientific Publishing Services, Chennai, India

Printed on acid-free paper

Springer is part of Springer Science+Business Media (www.springer.com)

Preface

Organizing the 4th International Workshop on Computational Topology in Image Context (CTIC 2012) has been an interesting experience indeed, going far beyond mathematics and computer science.

As mathematicians, we expected to focus only on theoretical problems and algorithms. On the contrary, the heaviest snowfall in Italy in the last 20 years reminded us that science is only part of reality, and forced the Organizing Committee to postpone the workshop for three months, until the end of May 2012. In one sense, this unpleasant occurrence was a reminder that unexpected phenomena are an important issue not only in scientific discovery but also in practical life.

This collection documents the presentations accepted at CTIC 2012. The research conducted by the authors of these papers was the core of the workshop, and we thank all the contributors for their commitment and dedication. Their effort allowed us to continue the tradition of CTIC in providing a forum for scientific exchange in topology and computation in image context at a high-quality level.

Special thanks go to our invited speakers, Frédéric Chazal (INRIA Saclay, Orsay) and Walter Kropatsch (PRIP, Vienna University of Technology), for their key contribution to the success of this workshop.

We also thank all the Scientific Committee members for their valuable feedback, which enabled the authors to further improve the quality of their work.

CTIC 2012 could not have been organized without our sponsors (Università degli Studi di Bologna, European Science Foundation, Rotary Club Bologna, GNSAGA—Istituto Nazionale di Alta Matematica "Francesco Severi") and supporters (Advanced Research Center on Electronic Systems for Information and Communication Technologies "E. De Castro"—University of Bologna, GIRPR—Gruppo Italiano Ricercatori in Pattern Recognition, SIMAI—Società Italiana di Matematica Applicata e Industriale, Università degli Studi di Modena e Reggio Emilia). We are very grateful to all of them.

We also thank the team working at the Centro Congressi in Bertinoro for their valuable help.

Finally, we are grateful to the participants attending this workshop, and to their snow-proof patience.

May 2012

Massimo Ferri
Patrizio Frosini
Claudia Landi
Andrea Cerri
Barbara Di Fabio

Organization

Scientific Committee Chairs

Massimo Ferri — Università di Bologna, Italy
Patrizio Frosini — Università di Bologna, Italy
Claudia Landi — Università di Modena e Reggio Emilia, Italy
Andrea Cerri — Università di Bologna, Italy
Barbara Di Fabio — Università di Bologna, Italy

Scientific Committee

Sylvie Alayrangues — Université de Poitiers, France
Antonio J. Bandera Rubio — Universidad de Málaga, Spain
Gilles Bertrand — PSI, ESIEE, Paris, France
Isabelle Bloch — CNRS LTCI Paris, France
Michael M. Bronstein — Università della Svizzera Italiana, Switzerland
Alex M. Bronstein — Tel Aviv University, Israel
David Cohen-Steiner — INRIA Sophia Antipolis, France
Michel Couprie — Université Paris-Est, France
Leila De Floriani — Università di Genova, Italy
Daniel Díaz-Pernil — Universidad de Sevilla, Spain
Herbert Edelsbrunner — Institute of Science and Technology of Austria
Bianca Falcidieno — IMATI-CNR Genova, Italy
Laurent Fuchs — Université de Poitiers, France
Edel B. García Reyes — CENATAV, Habana, Cuba
Antonio Giraldo Carbajo — Universidad Politécnica de Madrid, Spain
Rocío González-Díaz — Universidad de Sevilla, Spain
Yll Haxhimusa — PRIP, Vienna University of Technology, Austria
Adrian Ion — PRIP, Vienna University of Technology, Austria
María José Jiménez Rodríguez — Universidad de Sevilla, Spain
Tomasz Kaczynski — Université de Sherbrooke, Canada
Reinhard Klette — University of Auckland, New Zealand
Walter Kropatsch — PRIP, Vienna University of Technology, Austria
Jacques-Olivier Lachaud — Université de Savoie, France
Pascal Lienhardt — Université de Poitiers, France
Rémy Malgouyres — Université d'Auvergne, France
Jean-Luc Mari — Université de la Méditerranée, Marseille, France
Marian Mrozek — Jagiellonian University, Kraków, Poland

Darian Onchis-Moaca	Universität Wien, Austria
Nicolas Passat	Université de Strasbourg, France
Marcello Pelillo	Università Ca' Foscari, Venezia, Italy
Samuel Peltier	Université de Poitiers, France
Paweł Pilarczyk	Universidade do Minho, Portugal
Pedro Real Jurado	Universidad de Sevilla, Spain
Michela Spagnuolo	IMATI-CNR Genova, Italy
Peer Stelldinger	Universität Hamburg, Germany
José Antonio Vilches Alarcón	Universidad de Sevilla, Spain
Guy Wallet	Université de la Rochelle, France

Sponsoring Institutions

European Science Foundation (ESF), Strasbourg, France
Rotary Club Bologna, Italy
INdAM, Gruppo Nazionale per le Strutture Algebriche, Geometriche e le loro Applicazioni (GNSAGA), Italy
Alma Mater Studiorum, Università di Bologna, Italy
Università degli Studi di Modena e Reggio Emilia, Italy
Società Italiana di Matematica Applicata e Industriale (SIMAI), Italy
Gruppo Italiano Ricercatori in Pattern Recognition (GIRPR), Italy
Advanced Research Center on Electronic Systems for Information and Communication Technologies E. De Castro (ARCES), Università di Bologna, Italy

Table of Contents

A Framework for Label Images 1
   Loïc Mazo

Perfect Discrete Morse Functions on Triangulated 3-Manifolds 11
   Rafael Ayala, Desamparados Fernández-Ternero, and
   José Antonio Vilches

Removal Operations in nD Generalized Maps for Efficient Homology
Computation 20
   Guillaume Damiand, Rocio Gonzalez-Diaz, and Samuel Peltier

Enhancing the Reconstruction from Non-uniform Point Sets Using
Persistence Information 30
   Erald Vuçini

Parallel Skeletonizing of Digital Images by Using Cellular Automata.... 39
   Francisco Peña-Cantillana, Ainhoa Berciano,
   Daniel Díaz-Pernil, and Miguel A. Gutiérrez-Naranjo

Towards a Certified Computation of Homology Groups for Digital
Images 49
   Jónathan Heras, Maxime Dénès, Gadea Mata, Anders Mörtberg,
   María Poza, and Vincent Siles

An Efficient Algorithm to Compute Subsets of Points in $\mathbb{Z}^n$ 58
   Ana Pacheco and Pedro Real

Computational Topology in Text Mining 68
   Hubert Wagner, Paweł Dłotko, and Marian Mrozek

Concentrated Curvature for Mean Curvature Estimation in
Triangulated Surfaces 79
   Mohammed Mostefa Mesmoudi, Leila De Floriani, and Paola Magillo

Deletion of $(26,6)$-Simple Points as Multivalued Retractions 88
   Carmen Escribano, Antonio Giraldo, and María Asunción Sastre

Topological Operators on Cell Complexes in Arbitrary Dimensions 98
   Lidija Čomić and Leila De Floriani

Triangle Mesh Compression and Homological Spanning Forests 108
   Javier Carnero, Helena Molina-Abril, and Pedro Real

Homology Computations via Acyclic Subspace 117
   Piotr Brendel, Paweł Dłotko, Marian Mrozek, and Natalia Żelazna

Multi-scale Approximation of the Matching Distance for Shape
Retrieval 128
   Andrea Cerri, Barbara Di Fabio, and Filippo Medri

Persistent Homology for 3D Reconstruction Evaluation 139
   *Antonio Gutierrez, David Monaghan, María José Jiménez, and
   Noel E. O'Connor*

Persistence Modules, Shape Description, and Completeness 148
   *Francesca Cagliari, Massimo Ferri, Luciano Gualandri, and
   Claudia Landi*

Author Index 157

A Framework for Label Images

Loïc Mazo*

Université de Strasbourg, LSIIT, UMR CNRS 7005, France
loic.mazo@unistra.fr

Abstract. Label images need a specific topological model to take into account not only the topologies of the regions but also the topology of the partition. We propose a framework for label images in which all the regions of the initial partition and of any coarser partition of the space can be explicitly represented. Some properties of the model are given and a local transformation that preserves the weak homotopy types of all the regions of all the partitions is defined.

Keywords: Label image, simple point, homotopy type.

1 Introduction

In this paper, we study, from a topological viewpoint, *digital label images*, that is, images whose domain is $\mathbb{Z}^n$ and whose codomains are sets on which there generally exists no meaningful order relation (unlike grey-level images for instance). Label images need a specific approach in topology. Indeed, a label image is much more than a collection of independent objects and we are also interested in the spatial relations between these objects. Thereby, any topologically sound label image processing must pay attention to the objects *and* to the partition of the space associated to the labels. Nevertheless, as far as we know, the literature devoted to the topology of label images is not well developed and essentially oriented towards specific applications. The most commonly used approach is to process one label at a time while rejecting temporarily the other labels in the background, coming down to a binary image (*e.g.* [14,3,8]). With this method, either the topology of the partition is ignored or it is necessary to adjoin another structure, like a region adjacency graph, with the possibility to lose some information on tunnels or knots. Even in the case where the configuration permits a binary treatment of the image (in [9,12] the objects are concentric topological spheres), there are some specific issues. In a binary image, one usually finds just one object of interest embedded in an ambient space without significance. Hence, the choice of the adjacency pair can be done accordingly to the (known or expected) properties of the unique considered object. Now, if the binary image is a label image, that is to say, the *two* members of the partition are regions of interest, how to decide the region that will be equipped with the 6-adjacency and the

* The research leading to these results has received funding from the French *Agence Nationale de la Recherche* (Grant Agreement ANR-2010-BLAN-0205).

M. Ferri et al. (Eds.): CTIC 2012, LNCS 7309, pp. 1–10, 2012.

one that will be equipped with the 26-adjacency relation for instance? In [12], the choice is made in relation with the nature of the objects (their thickness) and in [9], the choice is made after algorithmic considerations. In both cases, that could be unsatisfactory if the two objects of interest are similar. Furthermore, most of the time, in the applications (see, *e.g.*, [13]) the value of the image on a picture element changes from the background to a particular label (depending on a cost, or energy, function), or *vice versa*, but more rarely it goes from a label to another label (and it occurs that points cannot be labelled since any choice of a label would break an *a priori* knowledge or would create a forbidden configuration like a cross over). An important drawback of this method is that an object identified by a label is inevitably seen under distinct adjacencies at distinct times of the process and even the number of the connected components is no more an invariant. This problem can be solved by working in the class of well-composed images [7] which uses the same adjacency relation for the object and for the background (namely, the $(2n)$-adjacency relation). Moreover, it is possible to transform (in a non-injective manner) an ordinary label image into a well-composed image [17] but the algorithm needs an order on the labels. The approach proposed in [4] makes it possible to change the label of a point in a 3D cellular space by choosing a new label among several ones with the assurance that the homotopy types of the two labels, the new one and the former one, are preserved. Nevertheless, no attention is paid to the topology of the partition. To take this latter topology in consideration, it is required in [2] that the unions of two labels (in 2D spaces), or two or three labels (in 3D spaces), are watched in any process as well as single labels. However, a careful examination of the examples provided by the authors shows that these conditions are not sufficient to maintain the topology of the partition. In particular, the unions of three labels should also be watched in a 2D space. In [5], the authors study label images in $\mathbb{Z}^3$ with objects that are 6-connected. They understand the topology of the partition as the set of the topologies of the surfaces between 6-adjacent labels. When performing a change of label on a point, they use simple points in the $(6, 18)$-adjacency framework of $\mathbb{Z}^3$ to maintain both topologies of the thinned label and of the growing one, while they use collapses in the cubical complex framework to ensure the preservation of the topologies of the surfaces between 6-adjacent labels. In a previous work [11], we have proposed an extensive theoretical framework to deal with label images. The main idea is that, topologically, a label image is a set of regions (that share the same label) together with some unions of these regions (thus, we retain the idea exposed in [2]). In other words, a label image is a partition together with some coarser partitions which are meaningful and a topologically sound processing of the image must control the topologies of all the regions of these partitions. In our framework, the label images are defined on a poset (partially ordered set), typically the space of cubical complexes equipped with the inclusion, and take their values on an atomistic[1] lattice of labels. The atoms of the lattice are the labels of the initial digital image (defined on $\mathbb{Z}^n$)

---

[1] A lattice is atomistic if any element, but the minimum, is a supremum of atoms.

and the supremum operation is used to create the unions of labels of interest (the coarser partitions). At least, besides the atoms, the lattice of labels must contain a minimum and a maximum (leading to the lattice of labels described in [15,16]). At most, the lattice is the power set of the atoms. In [11], we describe some transformations on label images that preserve the weak homotopy type of all the regions of interest. Nevertheless, this model has an important drawback at the first stage: the embedding of the digital space in a poset. If the digital objects are not modeled by closed subsets of $\mathbb{R}^n$ (*i.e.* if the adjacency relation for the objects is not the $(3^n - 1)$-adjacency), the embedding cannot preserve the topology of the labels together with their unions as illustrated on Figure 1. To overcome this issue, we have been lead to develop another model that we describe in this article.

(a)   (b)

Fig. 1. (a) A digital image λ in $\mathbb{Z}^2$ with 2 labels r (red), g (green) and a background (not depicted). (b) The embedding of the image λ in $\mathbb{F}^2$, the space of cubical 2-complexes, obtained by applying the following membership rule: the label of a face is the infimum of the labels of the surrounding facets in the poset $(2^{\{r,g\}}, \subseteq)$. We have proved in [10] that this embedding preserves the connected components and the fundamental groups of the object and its complement when a binary digital image is interpreted with the $(2n, 3^n - 1)$-adjacency pair. But, if we identify the two labels, that is if we consider a coarser partition of the space, the topology is not the same on these two images (we have one component on the left and two components on the right).

The remainder of this article is organized as follows. In Section 2, we define the covering images, the kind of abstract image that we propose to model a label digital image. In Section 3, we expose a notion of simple point for covering images. In Section 4, we show that the classical duality for binary images (object/background) can be extended to covering images. Section 5 concludes this paper and indicates some results, more technical, that we cannot expose here by lack of space. Note that, also by lack of space, no proofs are given in this article.

2   Covering Images

In order to model all the topological relations that can be found in a digital label image λ (defined on $\mathbb{Z}^n$), we propose two steps.

1. We split the image λ in a collection of binary images that represent the regions of interest, that is the regions that have been labeled (for instance during a segmentation process) and a number of unions of these regions

if we are interested in some inter-labels relations. The unions are labeled thanks to a lattice structure on labels: the label of an union of regions is the supremum of the labels of the regions. No other labels are needed for our purpose, so the lattice of labels, noted T, is an atomistic lattice (whose atoms are the initial labels) and there is a one-to-one correspondence between this lattice and the collection of binary images (we write λ_t for the binary image associated to the label t in the collection built from the digital label image λ). Figure 2 exemplifies this first stage (in the sequel, the infimum and supremum operators on T are denoted $\wedge$ and $\vee$ while $\perp$ and $\top$ are the minimum and the maximum of T).

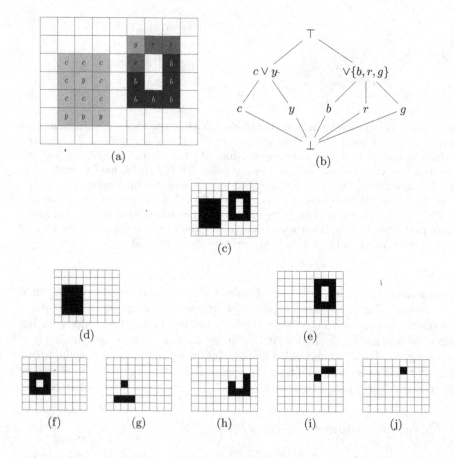

Fig. 2. (a) A digital label image with five labels c, y, b, r, g respectively depicted in cyan, yellow, blue, red and green. (b) A lattice structure T whose atoms are c, y, b, r, g. (c-j) The collection of binary images associated to the lattice T (the binary image associated to $\perp$ is not represented for it is a constant image, with no object).

2. In each binary digital image λ_t created in the previous step, we introduce inter-xels elements (pointels, linels, surfels and so on) to be able to use topological tools. The space in which we embed the digital images is the space of cubical complexes, $\mathbb{F}^n$, but it could be another cellular decomposition of the space[2]. The inter-xels elements must be labeled with membership rules with respect to the desired interpretation of the image. Such rules can be found in the literature (see *e.g.* [1]). We have proposed in [10] our own rules which preserve the connected components for the classical adjacency pairs and result in isomorphisms between the digital fundamental groups as defined by Kong and the classical fundamental groups of the regions of $\mathbb{F}^n$. Note that distinct rules can be applied to distinct labels provided no inconsistency is introduced (for instance, if two voxels are connected in the region labeled A, they cannot be disconnected in the region labeled by A or B): this leads us to the notion of fiber described below.

After these two steps have been achieved, we get a collection $(\mu_t)_{t\in T}$ (actually a lattice) of binary images, the *sheets*, defined on $\mathbb{F}^n$ (see Figure 3 and Figure 4). Now, we can see this collection as a unique image μ by setting that $\mu(x)$ is equal

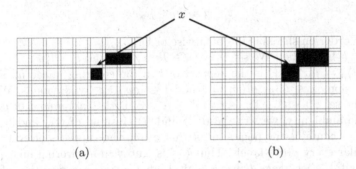

(a)   (b)

Fig. 3. The sheet associated to the label r of the digital label image depicted on Figure 2. (a) The digital binary image of Figure 2(i) is interpreted with the $(4, 8)$-adjacency pair (the object of the binary image is open, it does not include its boundary). (b) The same image is interpreted with the $(8, 4)$-adjacency pair (the object is closed so it includes its boundary).

to the set (we say the *fiber*) whose elements are the (extended) labels t such that $\mu_t(x) = 1$ (in other words, the labels attached to x, or, equivalently, the regions of interest x belongs to). For instance, let μ, resp. ν, be the collection of binary images a sheet of which is depicted on Figure 3(a), resp. Figure 3(b). Then, provided the same membership rule is applied on all the sheets, $\mu(x) = \{\vee\{b, r, g\}, \top\}$ while $\nu(x) = \{r, g, \vee\{b, r, g\}, \top\}$ where x is the one-dimensional face marked on Figure 3. When a point in $\mathbb{F}^n$ has not been labeled, for instance,

---

[2] A formal description of $\mathbb{F}^n$ can be found in [11] but no knowledge of cubical complexes is needed to understand the sequel of the article.

a point in the infinite region surrounding the image, then its fiber is set to $\emptyset$. Since the region obtained by identifying two labels a and b, that is, the region associated to the supremum of the labels a and b, contains all the faces that are in the region a or in the region b plus, possibly, some other faces (like the one-dimensional face at the center of Figure 1 when we take for a the red label and for b the green one), the fibers are *up-sets*, that is for any fiber S,

$$t \in S \text{ and } t \leq u \Rightarrow u \in S. \tag{1}$$

We write $\mathcal{G}_T$ for the family of the up-sets over the lattice T. Note that $\emptyset \in \mathcal{G}_T$.
Eventually, we can define a covering image.

Definition 1 (Covering image). *Let T be an atomistic lattice. A covering image μ is a function from $\mathbb{F}^n$ to $\mathcal{G}_T$. For any $t \in T$, the sheet μ_t is the binary image defined on $\mathbb{F}^n$ by $\mu_t(x) = 1$ if $t \in \mu(x)$ and $\mu_t(x) = 0$ otherwise. For any $x \in \mathbb{F}^n$, the set $\mu(x)$ is the fiber over x.*

From Definition 1 and Relation (1), we derive that, for any $t, u \in T$,

$$t \leq u \Rightarrow \mu_t \leq \mu_u. \tag{2}$$

where the order on sheets is the pointwise order.

Let $t \in T$ be a label. We set $t^\uparrow = \{u \in T \mid t \leq u\}$ and $t^\downarrow = \{u \in T \mid u \leq t\}$. The set $\langle t \rangle_\mu = \mu_t^{-1}(\{1\})$ is the *support* of t in the covering image μ. Thereby, the expressions $x \in \langle t \rangle_\mu$, $\mu_t(x) = 1$ and $t \in \mu(x)$ are synonymous. We write $\langle t \rangle_\mu^c$ for the set $\mathbb{F}^n \setminus \langle t \rangle_\mu$. The set $\langle T \rangle^c = \mu^{-1}(\emptyset)$ is the *background* of μ. When there is no ambiguity, we write also $\langle t \rangle$ and $\langle t \rangle^c$ instead of $\langle t \rangle_\mu$ and $\langle t \rangle_\mu^c$. The support $\langle \bot \rangle$ contains the points in $\mathbb{F}^n$ that are in the supports of every labels, in particular every proto-labels. Thus $\langle \bot \rangle$ is empty in a covering image obtain from a digital label image (since a digital label image is a partition of $\mathbb{Z}^n$, or a partition of a finite region of $\mathbb{Z}^n$). Nevertheless, we will need this support to define the dual of a covering image (Section 4): in the dual of a covering image built from a digital label image, the support $\langle \bot \rangle$ is not empty (but so is the background).

Since the lattice T is atomistic, and since, for any set of labels S, we have $(\bigvee_{t \in S} t)^\uparrow = \bigcap_{t \in S} t^\uparrow$, any element of the codomain $\mathcal{G}_T$ of a covering image can be expressed as an union of intersections :

$$\mathcal{G}_T = \left\{ \bigcup_{A \in \mathcal{B}} \bigcap_{t \in A} t^\uparrow \mid \mathcal{B} : \text{ a family of sets of atoms} \right\}$$

where $\bigcup_{A \in \emptyset} \bigcap_{t \in A} t^\uparrow = \emptyset$ and $\bigcap_{t \in \emptyset} t^\uparrow = \bot^\uparrow = T$. This latter description of $\mathcal{G}_T$ gives a way to encode the fibers with trees.

3 Simplicity

A more detailed presentation of the notions of topology used below can be found in our previous work on label images [11].

In a binary image, a point x in the object is *simple* if its removal from the object "preserves topology" [6]. Since a covering image is a collection of binary images (the sheets), we can extend the notion of simple point to covering images: roughly speaking, in a covering image, a point is simple for a fiber S if it is simple in any sheet modified by the assignment $\mu(x) = S$. In our framework, we use β-simple points. A point $x \in \mathbb{F}^n$ is a β-*simple point* for a subset X of $\mathbb{F}^n$ if one of the sets $x^\uparrow \setminus \{x\}$ or $x^\downarrow \setminus \{x\}$ is contractible[3]. The β-simple points have the advantage to preserve topology twice. Indeed, in the one hand, $\mathbb{F}^n$ can be equipped with the Alexandroff topology whose open sets are the up-sets of $\mathbb{F}^n$. Then, the deletion of a β-simple point x from a subset X of $\mathbb{F}^n$ is a weak homotopy equivalence, that is, the inclusion $i : X \setminus \{x\} \to X$ induces a one-to-one correspondence between the connected components of both spaces and induces also isomorphisms between the homotopy groups of $X \setminus \{x\}$ and X. Moreover, this is also true for the dual inclusion $i' : \mathbb{F}^n \setminus X \to \mathbb{F}^n \setminus (X \setminus \{x\})$. On the other hand, one can associate to any subset X of $\mathbb{F}^n$ an euclidean set, denoted $|\mathcal{K}(X)|$, which is the realization of a simplicial complex. The deletion from X of a β-simple point x induces a strong deformation retraction from $|\mathcal{K}(X)|$ to $|\mathcal{K}(X \setminus \{x\})|$. Furthermore, this is also true for the complements (in $\mathbb{R}^n$) of these realizations but, possibly, in a non-monotonic manner.

Eventually, we can give the definition of a simple point in a covering image.

Definition 2 (Simple point in a covering image). *Let $S \in \mathcal{G}_T$ be a fiber. A point $x \in \mathbb{F}^n$ is* simple for (the fiber) S *if the following two conditions are verified:*

   (i) for any label $u \in \mu(x)$ such that $u \notin S$, x is β-simple for the set $\langle u \rangle$ or for the set $\langle u \rangle^c \cup \{x\}$;

   (ii) for any label $u \notin \mu(x)$ such that $u \in S$, x is β-simple for the set $\langle u \rangle \cup \{x\}$ or for the set $\langle u \rangle^c$.

The previous definition, and the properties of β-simple points, ensures that, if a point is simple for the fiber S in the image μ, we can set $\mu(x) = S$ while preserving the topology of any region of interest, including the unions pointed out by the choice of the lattice T. Moreover, modifications of fibers over points in $\mathbb{F}^n$ having the same dimension can be done in parallel, leading to well-balanced algorithms. Figure 4 gives an example of thinning on a covering image using simple points for the fiber $\emptyset$. Observe that in this thinning we have maintained the possibility to go back to $\mathbb{Z}^n$ without losing any topological information but this is a difficult issue on which we are still working.

---

[3] A space is *contractible* if it has the homotopy type of a point. A subset X of $\mathbb{F}^n$ is contractible iff it can be shrunk to a unique point by the sequential removal of unipolar points. A point x is *unipolar* if $x^\uparrow \setminus \{x\}$ has a minimum or $x^\downarrow \setminus \{x\}$ has a maximum.

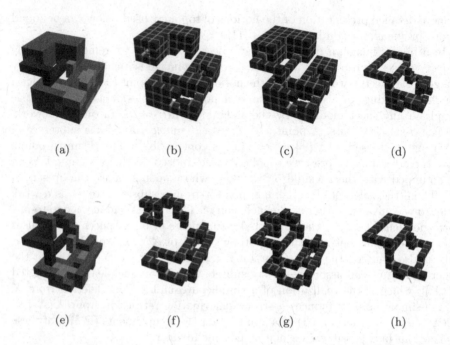

Fig. 4. (a) A label image λ defined on $\mathbb{Z}^3$. (b) The "red" sheet μ_{red} of the covering image μ associated to the image λ and the $(6, 18)$-adjacency pair (on this sheet, which is a binary image, as on the other sheets, the colors are used only to better distinguish the faces of the cubical complex). For any label t, and any $f \in \mathbb{F}^n$ such that $\dim(f) \leq 2$, the rule used to define the value of $\mu_t(f)$ can be expressed as follows: if $\dim(f) = d$ and f bounds at least k black voxels then f is black (*i.e.* $\mu_t(f) = 1$), otherwise f is white, where $(d, f) \in \{(2, 2), (1, 3), (0, 6)\}$ (see [10]). (c) The "red-or-blue" sheet. (d) The "green-or-blue" sheet. (e–h) The same as (a–d) after a thinning procedure (using simple points for the fiber $\emptyset$) has been applied. Note that in our implementation we do not directly process the sheets but the fibers (and we extract the sheets for visualization when the process is over).

4   Duality

Thanks to the dual order on the lattice T and to the complementation in 2^T, we are able to define the dual of a covering image.

We write $\mathcal{F}_T$ for the subset of 2^T composed of the complements of the elements of $\mathcal{G}_T$ in 2^T. As $\mathcal{G}_T$ is the family of the up-sets over T, $\mathcal{F}_T$ is the family of the down-sets over T ($E \in \mathcal{F}_T$ iff $e \in E$ and $f \leq e$ imply $f \in E$), *i.e.* $\mathcal{F}_T = \mathcal{G}_{(T^*)}$ where T^* is the lattice T equipped with its dual order.

Definition 3 (Dual covering image). *The dual covering image of μ is the covering image $\neg\mu : \mathbb{F}^n \to \mathcal{F}_T$ defined for any $x \in \mathbb{F}^n$ by $(\neg\mu)(x) = 2^T \setminus \mu(x)$.*

The duality on covering images comes down to a swap of the background and the object in each sheet of a covering image: for any label $t \in T$, $\langle t \rangle_{\neg\mu} = \langle t \rangle_\mu^c$ and the background of the dual of a covering image μ is $\mu^{-1}(T) = \mu^{-1}(\perp^\uparrow)$.

The duality is compatible with the simplicity: a face is simple for a fiber S in a covering image iff it is simple in the dual covering image for the complement in 2^T of S.

Thanks to this duality we can design in one time tools to process digital label images whose regions have to be interpreted with an (α, β)-adjacency pair or with the (β, α)-adjacency pair.

5 Conclusion

The model exposed in this paper is a way to encompass all the topological relations that characterize a label image. From a theoretical point of view, it can help us to check what is precisely preserved or modified by a procedure. Likely, it is not necessary to implement the model as is. We have begun to look for sufficient conditions to directly process label images defined on $\mathbb{Z}^3$ with the conceptual aid of covering images. The first results are tedious but we have not made any optimization on them. On the other hand, it seems predictable that the complexity of a label image (with all the intra-labels and inter-labels relations) is exponentially more costly than the complexity of a single object. Nevertheless, we think that this model can be useful to work on images with several objects when the inter-relations between the objects are meaningful.

Further works will consist in improving the implementation of the whole framework and testing this implementation on "real" images. We plan also to define and study morphological operators for covering images.

References

1. Ayala, R., Domínguez, E., Francés, A.R., Quintero, A.: Digital Lighting Functions. In: Ahronovitz, E. (ed.) DGCI 1997. LNCS, vol. 1347, pp. 139–150. Springer, Heidelberg (1997)
2. Bazin, P.-L., Ellingsen, L.M., Pham, D.L.: Digital Homeomorphisms in Deformable Registration. In: Karssemeijer, N., Lelieveldt, B. (eds.) IPMI 2007. LNCS, vol. 4584, pp. 211–222. Springer, Heidelberg (2007)
3. Bazin, P.-L., Pham, D.L.: Topology-preserving tissue classification of magnetic resonance brain images. IEEE Transactions on Medical Imaging 26(4), 487–496 (2007)
4. Cointepas, Y., Bloch, I., Garnero, L.: A cellular model for multi-objects multi-dimensional homotopic deformations. Pattern Recognition 34, 1785–1798 (2001)
5. Damiand, G., Dupas, A., Lachaud, J.-O.: Fully deformable 3D digital partition model with topological control. Pattern Recognition Letters 32, 1374–1383 (2011)

6. Kong, T.Y., Rosenfeld, A.: Digital topology: Introduction and survey. Computer Vision, Graphics, and Image Processing 48, 357–393 (1989)
7. Latecki, L.J.: Multicolor well-composed pictures. Pattern Recognition Letters 16(4), 425–431 (1995)
8. Liu, J., Huang, S., Nowinski, W.: Registration of brain atlas to MR images using topology preserving front propagation. Journal of Signal Processing Systems 55(1), 209–216 (2009)
9. Mangin, J.-F., Frouin, V., Bloch, I., Régis, J., López-Krahe, J.: From 3D magnetic resonance images to structural representations of the cortex topography using topology preserving deformations. Journal of Mathematical Imaging and Vision 5(4), 297–318 (1995)
10. Mazo, L., Passat, N., Couprie, M., Ronse, C.: Digital imaging: A unified topological framework. Journal of Mathematical Imaging and Vision (to appear, 2012), doi:10.1007/s10851-011-0308-9
11. Mazo, L., Passat, N., Couprie, M., Ronse, C.: Topology on digital label images. Journal of Mathematical Imaging and Vision (to appear, 2012), doi:10.1007/s10851-011-0325-8
12. Miri, S., Passat, N., Armspach, J.-P.: Topology-Preserving Discrete Deformable Model: Application to Multi-segmentation of Brain MRI. In: Elmoataz, A., Lezoray, O., Nouboud, F., Mammass, D. (eds.) ICISP 2008. LNCS, vol. 5099, pp. 67–75. Springer, Heidelberg (2008)
13. Pham, D., Bazin, P.-L., Prince, J.: Digital topology in brain imaging. IEEE Signal Processing Magazine 27(4), 51–59 (2010)
14. Poupon, F., Mangin, J.-F., Hasboun, D., Poupon, C., Magnin, I.E., Frouin, V.: Multi-object Deformable Templates Dedicated to the Segmentation of Brain Deep Structures. In: Wells, W.M., Colchester, A.C.F., Delp, S.L. (eds.) MICCAI 1998. LNCS, vol. 1496, pp. 1134–1143. Springer, Heidelberg (1998)
15. Ronse, C., Agnus, V.: Morphology on label images: Flat-type operators and connections. Journal of Mathematical Imaging and Vision 22(2), 283–307 (2005)
16. Ronse, C., Agnus, V.: Geodesy on label images, and applications to video sequence processing. Journal of Visual Communication and Image Representation 19, 392–408 (2008)
17. Siqueira, S., Latecki, L.J., Tustison, N., Gallier, J., Gee, J.: Topological repairing of 3D digital images. Journal of Mathematical Imaging and Vision 30(3), 249–274 (2008)

Perfect Discrete Morse Functions on Triangulated 3-Manifolds

Rafael Ayala, Desamparados Fernández-Ternero, and José Antonio Vilches

Departamento de Geometría y Topología, Universidad de Sevilla,
P.O. Box 1160 41080 Sevilla Spain,
{rdayala,desamfer,vilches}@us.es

Abstract. This work is focused on characterizing the existence of a perfect discrete Morse function on a triangulated 3-manifold M, that is, a discrete Morse function satisfying that the numbers of critical simplices coincide with the corresponding Betti numbers. We reduce this problem to the existence of such kind of function on a spine L of M, that is, a 2-subcomplex L such that $M - \Delta$ collapses to L, where Δ is a tetrahedron of M. Also, considering the decomposition of every 3-manifold into prime factors, we prove that if every prime factor of M admits a perfect discrete Morse function, then M admits such kind of function.

Keywords: perfect discrete Morse function, triangulated 3-manifold, spine.

1 Introduction

The existence of a perfect discrete Morse function on a triangulated 3-manifold K heavily depends on K and gives us information about its combinatorial structure. For example, it is easy to prove that if we consider a triangulation of S^3 containing the Dunce hat as subcomplex then any discrete Morse function defined on it has at least 3 critical simplices. This situation strongly contrasts with the smooth case, where the existence of perfect functions on a manifold does not depend on the given triangulation. The main goal of this work, which is the continuation of the paper [2], consists on reducing the problem, initially stated on 3-manifolds, to 2-complexes. It is possible since we prove that a triangulation K of a 3-manifold admits a perfect discrete Morse function if and only if a spine of K admits such kind of function. For the sake of simplicity, we have restricted our study to closed orientable 3-manifolds.

This problem can be regarded not only in a theoretical way but also it is strongly linked with applications in several areas like digital image processing, object recognition and representation of digital objects. In this sense there are works which use 3-dimensional cell complexes for modelling $4D$ digital objects, which arise in a natural way when considering a time sequence of $3D$ objects (see [19] and [5]). Any discrete Morse function provides information on the homology of a manifold. For example, the number of critical i-cells is not less than the i-th Betti number of the manifold. In this context, a perfect discrete Morse

M. Ferri et al. (Eds.): CTIC 2012, LNCS 7309, pp. 11–19, 2012.
© Springer-Verlag Berlin Heidelberg 2012

function on a complex encodes its minimal topological structure in terms of the basic n-cycles. Notice that it is not always possible to get a perfect discrete Morse function, since there are complexes not admitting such kind of functions. For example, there are triangulations of 3-balls containing complicated knots as subcomplexes with few edges not admitting perfect functions (see [4] for details). Optimal discrete Morse functions (those with the minimal number of critical cells) can be used to obtain minimal decompositions of complexes/digital images in terms of homological forests (see [17] for details). Only in the case that the considered complex admits perfect discrete Morse function, such minimal decomposition has as many critical i-cells as the i-th Betti number and, moreover, the triangulation of the model is "nice", that is, the complex contains no complicated knot (see [4]).

Discrete Morse functions arise in a natural way in the image context as graylevel scale digital images are taken into account. Being more precise, starting from real values (graylevel scale) on the vertices of a cubical complex, they are extended to the complex by defining a discrete Morse function on it (see [21] and [11]).

The paper is organized as follows: Section 2 is devoted to introducing the basic notions and results concerning to discrete Morse theory. Section 3 includes the obtained results on graphs and 2-complexes. Once the main result is obtained at the beginning of Section 4, that is, the reduction of the problem of deciding if a given triangulated 3-manifold admits perfect discrete Morse functions to the 2-dimensional case, we indicate necessary conditions in terms of the homology of the manifold for the existence of perfect functions. Also, taking into account the decomposition of a 3-manifold into prime manifolds, we prove in a constructive way that if every prime factor admits a perfect function then the given manifold admits such a kind of function.

2   Preliminaries

Through all this paper we shall consider finite simplicial complexes.

Let K be a simplicial n-complex and α be a n-simplex of K. If there exists a $(n-1)$-dimensional face β of α such that β is not a face of any other n-simplex in K, we say that there is an **elementary collapse** from K to $K - \{\alpha, \beta\}$. The inverse operation is called an **elementary expansion** from $K - \{\alpha, \beta\}$ to K. If $K = K^0 \supset K^1 \supset \cdots \supset K^m = L$ are simplicial complexes such that there is an elementary collapse from K^{i-1} to K^i, $i = 1, ..., m$, we say that K **collapses to** L, denoted by $K \searrow L$. Equivalently, the inverse operation is called an **expansion** from L to K, denoted by $K \nearrow L$. The notion of collapse was introduced by J.C.H. Whitehead [22] in the context of simple homotopy theory. More recently, there have been introduced weaker notions like *shaving* in the computation of the homology of cubical complexes [18] and *contraction* in the computation of cohomology operations [10] which are more suitable for Computational Topology.

The **collapse number** of a 2-complex K, denoted by $co(K)$, is the minimal number of 2-simplices $\tau_1, \ldots, \tau_{co(K)}$ that need to be removed from K so that $K - \{\tau_1, \ldots, \tau_{co(K)}\}$ collapses to a graph. A detailed study of the computability of this number can be found in [6].

A 2-complex such that every 1-simplex is a face of exactly two 2-simplices is called a **2-pseudomanifold**. A 2-complex K is said to be **strongly connected** if, given any two 2-simplices σ, σ' in K, there exists a chain of 2-simplices connecting them, that is, a sequence $\sigma_0, \ldots, \sigma_k$ of 2-simplices such that $\sigma = \sigma_0$, $\sigma' = \sigma_k$ and $\sigma_i \cap \sigma_{i+1}$ is a common 1-face. The **strongly connected components** of a 2-complex are its maximal strongly connected subcomplexes. A strongly connected 2-pseudomanifold is called a **strong 2-pseudomanifold**. Notice that any 2-pseudomanifold K can be decomposed as the union of its strongly connected components and the intersection of two of such components is either empty or a finite set of vertices.

Given a simplicial complex K, a **discrete Morse function** is a function $f : K \longrightarrow \mathbb{R}$ such that, for any p-simplex $\sigma \in K$:

(M1) $card\{\tau^{(p+1)} > \sigma / f(\tau) \leq f(\sigma)\} \leq 1.$
(M2) $card\{v^{(p-1)} < \sigma / f(v) \geq f(\sigma)\} \leq 1.$

A p-simplex $\sigma \in K$ is said to be **a critical simplex** with respect to f if:

(C1) $card\{\tau^{(p+1)} > \sigma / f(\tau) \leq f(\sigma)\} = 0.$
(C2) $card\{v^{(p-1)} < \sigma / f(v) \geq f(\sigma)\} = 0.$

A value of a discrete Morse function on a critical simplex is called **critical value**.

Given $c \in \mathbb{R}$, the **level subcomplex** $K(c)$ is the subcomplex of K consisting of all simplices τ with $f(\tau) \leq c$, as well as all of their faces, that is,

$$K(c) = \bigcup_{f(\tau) \leq c} \bigcup_{\sigma \leq \tau} \sigma$$

Given two values of f, $a_k < a_l$, the relationship between two level subcomplexes $K(a_k)$ and $K(a_l)$ is the following [7]:

If the interval $[a_k, a_l]$ does not contain any critical value, then $K(a_l)$ collapses to $K(a_k)$ or equivalently, $K(a_k)$ expands to $K(a_l)$.

If the interval $[a_k, a_l]$ contains a critical value corresponding to a critical simplex of dimension i, then $K(a_l)$ has the same simple homotopy type as $K(a_k)$ with an i-cell attached.

A **discrete vector field** V on K is a collection of pairs $(\alpha^{(p)} < \beta^{(p+1)})$ of simplices of K such that each simplex is in at most one pair of V. A **V-path** is a sequence of simplices

$$\alpha_0^{(p)}, \beta_0^{(p+1)}, \alpha_1^{(p)}, \beta_1^{(p+1)}, \ldots, \beta_r^{(p+1)}, \alpha_{r+1}^{(p)}, \ldots,$$

such that, for each $i \geq 0$, the pair $(\alpha_i^{(p)} < \beta_i^{(p+1)}) \in V$ and $\beta_i^{(p+1)} > \alpha_{i+1}^{(p)} \neq \alpha_i^{(p)}$.

Given a discrete Morse function f on K, the **gradient vector field** induced by f is the set of pairs of simplices $(\alpha^{(p)} < \beta^{(p+1)})$ such that $f(\alpha) \geq f(\beta)$.

Theorem 1. *[8] A discrete vector field V is the gradient vector field of a discrete Morse function if and only if there are no non-trivial closed V-paths.*

Theorem 2. *[7] Let f be a discrete Morse function defined on K and let b_p be the p-th Betti number of X with $p = 0, 1, \ldots, n$ (where n is the dimension of K). Then:*

(I1) $m_p(f) - m_{p-1}(f) + \cdots + (-1)^p m_0 \geq b_p - b_{p-1} + \cdots + (-1)^p b_0$,
(I2) $m_p(f) \geq b_p$,
(I3) $m_0(f) - m_1(f) + \cdots + (-1)^n m_n(f) = b_0 - b_1 + \cdots + (-1)^n b_n = \chi(X)$,

where $m_p(f)$ denotes the number of critical p-simplices of f on K.

Notice that these inequalities are still valid for the case of Betti numbers with general coefficients, that is, using any field F instead of $\mathbb{Z}$.

In the smooth setting Pitcher proved in [20] the following generalized version of Morse inequalities which also take into account the torsion coefficients q_p of $H_p(M; F)$, where F is a field or $\mathbb{Z}$:

Theorem 3. *Let M be a compact Riemannian manifold and let f be a smooth Morse function. Then:*

1. $m_0(f) \geq b_0$, $m_p(f) \geq b_p + q_p + q_{p-1}$ with $p = 1, \ldots, n-1$, $m_n(f) \geq b_n + q_{n-1}$.
2. $m_p(f) - m_{p-1}(f) + \cdots + (-1)^p m_0 \geq b_p - b_{p-1} + \cdots + (-1)^p b_0 + q_p$ with $p = 1, \ldots, n-1$.

This result can be extended in a straightforward way to the discrete approach.

A discrete Morse function f defined on K is **optimal** if it has the least possible number of critical simplices, that is, $m_i(f) \leq m_i(g)$ with $1 \leq i \leq n$ for every discrete Morse function g on K.

A discrete Morse function f is called F-**perfect** if $m_p(f) = b_p(K; F)$ with $p = 0, \ldots, n$. Taking into account the Morse inequality (I2) of Theorem 2, we conclude that every F-perfect function is optimal.

Remark 1. Notice that if a discrete Morse function is $\mathbb{Z}$-perfect then, due to $\beta_i(K, \mathbb{Z}) = \beta_i(K, \mathbb{Q})$, the function is $\mathbb{Q}$-perfect too. Moreover, if a complex K admits a $\mathbb{Q}$-perfect discrete Morse function, then the homology groups of K with coefficients in $\mathbb{Z}$ are torsion-free (see Proposition 5.9 of [14] for details). By means of the universal coefficient theorem, the Betti numbers are the same for coefficients in any field and consequently the function is F-perfect for all field F.

3 Perfect Discrete Morse Functions on Graphs and 2-Complexes

We will start the study of the existence of perfect discrete Morse functions on 2-complexes by considering the case of homology with integer coefficients. It is well known that every 1-dimensional complex admits $\mathbb{Z}$-perfect discrete Morse functions. It can be proved considering a spanning tree, by means of Lemma 4.3 of [7], a discrete Morse function on such tree with a unique critical vertex is constructed. Finally, this function is extended to the graph by assigning a local maximum to every edge not contained in the tree.

As a direct consequence, taking into account Lemma 4.3 of [7], every 2-complex collapsing to a graph admits a $\mathbb{Z}$-perfect discrete Morse function. In particular, since every surface with boundary collapses to a graph (see example 7, page 52 of [9]), such surfaces admit a $\mathbb{Z}$-perfect discrete Morse function.

Now we are going to give several results on the links between the existence of $\mathbb{Z}$-perfect discrete Morse functions on a given 2-complex with some trivial homology groups and its simple homotopy type.

Proposition 1. *Let K be a compact connected 2-complex admitting a $\mathbb{Z}$-perfect discrete Morse function. The following statements hold:*

1. *If K is $\mathbb{Z}$-acyclic then it is collapsible.*
2. *If $H_1(K) = 0$ and $H_2(K) \neq 0$ then K has the same simple homotopy type as a wedge of copies of S^2.*
3. *If $H_1(K) \neq 0$ and $H_2(K) = 0$ then K has the same simple homotopy type as a wedge of copies of S^1.*

It follows from the above result that every compact connected 2-complex which is $\mathbb{Z}$-acyclic and non-contractible does not admit $\mathbb{Z}$-perfect discrete Morse functions. An example of such kind of 2-complex can be found in [16].

Using a straightforward Mayer-Vietoris argument, we can prove that every complex K admitting $\mathbb{Z}$-perfect discrete Morse functions satisfies $H_1(K; \mathbb{Z})$ is free. Notice that the converse is not true. In [2] an example of a 2-complex K with $H_1(K) = \mathbb{Z}$ and not admitting $\mathbb{Z}$-perfect discrete Morse functions is included. In particular, if the first fundamental group of a 2-complex is finite and non-trivial then it does not admit $\mathbb{Z}$-perfect discrete Morse functions.

By means of the collapse number of a 2-complex we proved that a connected compact 2-complex K admits a $\mathbb{Z}$-perfect discrete Morse function if and only if $co(K) = b_2(K; \mathbb{Z})$. As a direct consequence we obtained that a compact connected surface without boundary admits a $\mathbb{Z}$-perfect discrete Morse function if and only if it is orientable. Moreover, this result can be extended to 2-pseudomanifolds in the sense that a 2-pseudomanifold K admits a $\mathbb{Z}$-perfect discrete Morse function if and only if every strongly connected component of K is orientable.

4   Perfect Discrete Morse Functions on 3-Manifolds

In this section we will consider the problem of the existence of perfect discrete Morse functions on 3-manifolds. First of all, let us recall some basic notions and results concerning 3-manifolds.

Definition 1. *Given two closed triangulated 3-manifolds K and L, the* **connected sum** *of them, denoted by $K \sharp L$, is defined as follows:*

- *We choose two 3-simplices σ and τ in K and L respectively.*
- *We identify in $(K - \sigma) \cup (L - \tau)$ the simplices of the boundaries of σ and τ by some simplicial gluing map.*

Definition 2. *A connected triangulated 3-manifold M is called* **prime** *if $M = K \sharp L$ implies $K = S^3$ or $L = S^3$.*

Theorem 4 (Knesser-Milnor[12]). *Let M be a compact connected orientable triangulated 3-manifold. Then there is a decomposition $M = P_1 \sharp \cdots \sharp P_n$ with each P_i prime. This decomposition is unique up to insertion or deletion of S^3.*

Using the Pitcher's strengthened version of Morse inequalities we get the next result

Proposition 2. *Let M be a closed triangulated 3-manifold admitting a $\mathbb{Z}$-perfect discrete Morse function, then M is orientable and $H_1(M; F)$ is free for all field F.*

Remark 2. The converse of the above result is not true since there are 3-manifolds with $H_1(M)$ free which do not admit $\mathbb{Z}$-perfect discrete Morse functions as it is shown in the following example:

Consider the 3-manifold $M = H^3 \sharp (S^1 \times S^2)$, where H^3 denotes the Poincare's homology sphere (see [12] and [13]). Notice that $H_1(M) = \mathbb{Z}$ is free (in fact $H_i(M) = \mathbb{Z}$) and thus a perfect discrete Morse function has 4 critical simplices. However, since the first homotopy group $\Pi_1(M)$ is not abelian it follows that any presentation of this group has at least two generators and then the 1-skeleton of any CW-structure of M contains the wedge of at least two circles. So $m_2 = m_1 \geq 2$ and hence we conclude that every discrete Morse function on M has at least 6 critical simplices.

Corollary 1. *Let M be a closed triangulated 3-manifold such that $\Pi_1(M)$ is finite and non-trivial. Then M does not admit $\mathbb{Z}$-perfect discrete Morse functions.*

Remark 3. Examples of this kind of spaces are the spherical manifolds S^3/Γ where Γ is a finite subgroup of $SO(4)$ acting freely on S^3 by rotations. In the particular case that Γ is a cyclic group these manifolds are the lens spaces $L(p, q)$ for $0 < \frac{p}{q} < 1$.

Corollary 2. *Let M be a closed triangulated 3-manifold such that $\Pi_1(M)$ contains a torsion subgroup. Then M does not admit $\mathbb{Z}$-perfect discrete Morse functions.*

Definition 3. *Let M be a triangulated 3-manifold. A* **spine** *L of M is a 2-subcomplex L such that $M - \Delta$ collapses to L, where Δ is a tetrahedron of M.*

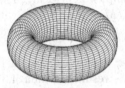

Fig. 1. The figure on the left is solid torus admitting as spine the torus on the right

The next figure represents a spine of the 3-manifold obtained by the complement in a 3-ball of a knotted thickened Y, the so called worm-eaten apple.

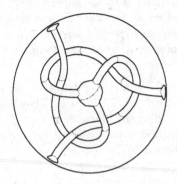

Fig. 2. A spine of a worm-eaten apple

Next result illustrates how the problem of deciding whether a given orientable 3-manifold M admits a $\mathbb{Z}$-perfect discrete Morse function is reduced to the two-dimensional case, that is, it is equivalent to the problem of determining the existence of such kind of functions on a spine of M. Algorithms for the obtention of perfect functions on 2-complexes are presented in [3].

Theorem 5. *Let M be a connected closed orientable triangulated 3-manifold and let F be either $\mathbb{Z}$ or a field. M admits an F-perfect discrete Morse function if and only if there exists L, a spine of M, which admits an F-perfect discrete Morse function.*

Proof. Let us suppose that M admits an F-perfect discrete Morse function f, hence $m_i(f) = b_i$ with $i = 0, \ldots, 3$. Since $H_3(M) \simeq \mathbb{Z}$ and M is connected, then $m_0(f) = 1 = m_3(f)$. By means of Poincare duality's theorem, we get that $m_1(f) = m_2(f)$. Let N be the subcomplex obtained by removing the unique critical 3-simplex of f denoted by Δ. Notice that N is of the same simple homotopy type as a 2-dimensional subcomplex L, since N can be obtained from L by a sequence of elementary expansions, in fact L is a spine of M. The proof can be carried out by induction on the sequence of level subcomplexes taking into account that every 3-simplex added at a given step provides us an expansion (since no 3-simplex is critical). Then, by an easy Mayer-Vietoris argument we get $b_i(L) = b_i(M - \Delta) = b_i(M)$ with $i = 0, 1, 2$. Indeed the restriction g of f to L is perfect satisfying $m_i(f) = m_i(g)$ with $i = 0, 1, 2$.

Conversely, let us assume that there is a spine L admitting an F-perfect discrete Morse function g. Let Δ be a 3-simplex such that $M - \Delta \searrow L$ then by a Forman's result (Lemma 4.3 of [7]), g can be extended to a $\mathbb{Z}$-perfect discrete Morse function $\widehat{g}$ on $M - \Delta$. Finally, we extend $\widehat{g}$ to M by defining $f(\Delta) = 1 + \max_{\sigma < \Delta} f(\sigma)$.

The following result guarantees the existence of a perfect discrete Morse function on a closed 3-manifold if every single component of its prime factor (given by Theorem 4) admits such a kind of function.

Theorem 6. *Let $M = M_1 \sharp M_2$ be a decomposition of a connected closed orientable 3-manifold M and let F be either $\mathbb{Z}$ or a field. If there exist triangulations K_i of M_i admitting an F-perfect discrete Morse function with $i = 1, 2$ then there exists a triangulation K of M admitting an F-perfect discrete Morse function.*

Proof. Let us assume that K_i is a triangulation of M_i admitting an F-perfect discrete Morse function with $i = 1, 2$. By using Theorem 5, there are spines L_i of K_i admitting F-perfect discrete Morse functions f and g. Let $L_1 \vee L_2$ be the wedge obtained by identifying the only critical vertices v and w of f and g respectively. Then the function h, defined as

$$h(\sigma) = \begin{cases} f(\sigma), & \text{if } \sigma \in L_1 - \{v\}; \\ g(\sigma), & \text{if } \sigma \in L_2 - \{w\}; \\ min\{f(v), g(w)\}, & \text{if } \sigma = v = w. \end{cases}$$

is an F-perfect discrete Morse function on $L_1 \vee L_2$.

Since it is known that $L_1 \vee L_2$ is a spine of $K = K_1 \sharp K_2$, (see [15]) then by means Theorem 5, we conclude that K (which is a triangulation of M) admits an F-perfect discrete Morse function.

Remark 4. Taking into account the classification of closed prime 3-manifolds based on their fundamental group (see [13]), by Corollary 1 we obtain that a closed prime 3-manifold M admitting a perfect discrete Morse function is either $S^1 \times S^2$ or $K(\Pi_1(M), 1)$ where $\Pi_1(M)$ is infinite but not cyclic.

In this second case, by means of Corollary 2 we conclude that $\Pi_1(M)$ is torsion free and also, if we suppose that $\Pi_1(M)$ is abelian then $M = S^1 \times S^1 \times S^1$. Now, since $H_1(M)$ is infinite then M must be a sufficiently large Haken manifold, for example, $M_g \times S^1$ where M_g is an orientable surface of genus g (see [13]).

References

1. Ayala, R., Fernández, L.M., Vilches, J.A.: Discrete Morse inequalities on infinite graphs. Electron. J. Comb. 16 (1), paper R38, 11 (2009)
2. Ayala, R., Fernández-Ternero, D., Vilches, J.A.: Perfect discrete Morse functions on 2- complexes. Pattern Recognition Lett. (2011), doi: 10.1016/j.patrec.2011.08.011

3. Ayala, R., Fernández-Ternero, D., Vilches, J.A.: Constructing optimal discrete Morse functions on certain 2-complexes (submitted)
4. Benedetti, B.: Discrete Morse Theory is as perfect as Morse Theory. arXiv:1010.0548v3 [math.DG]
5. Couprie, M., Bertrand, G.: New characterizations of simple points in 2D, 3D and 4D discrete spaces. IEEE Trans. Pattern Anal. Mach. Intell. 31(4), 637–648 (2009)
6. Eğecioğlu, Ö., Gonzalez, T.: A computationally intractable problem on simplicial complexes. Comput. Geom. 6(2), 85–98 (1996)
7. Forman, R.: Morse Theory for cell complexes. Adv. Math. 134(1), 90–145 (1998)
8. Forman, R.: A user's guide to discrete Morse theory. Sém. Lothar. Combin. 48, Art. B48c, 35 (2002) (electronic)
9. Glaser, L.C.: Geometrical Combinatorial Topology. Van Nostrand Reinhold Company, New York (1970)
10. González-Díaz, R., Real, P.: Computation of cohomology operations on finite simplicial complexes. Homology Homotopy Appl. 5(2), 83–93 (2003)
11. Gyulassy, A., Bremer, P.T., Hamann, B., Pascucci, V.: A practical approach to Morse-Smale complex computation: Scalability and generality. IEEE Trans. Vis. Comput. Graph. 14(6), 1619–1626 (2008)
12. Hatcher, A.: Notes on basic 3-manifold Topology, http://www.math.cornell.edu/~hatcher/3M/3Mdownloads.html
13. Hatcher, A.: The clasification of 3-manifolds. A brief overview, http://www.math.cornell.edu/~hatcher/Papers/3Msurvey.pdf
14. Jonsson, J.: Simplicial complexes of graphs. Lecture Notes in Math., vol. 1928. Springer, Berlin (2008)
15. Knutson, G.W.: A characterization of closed 3-manifolds with spines containing no wild arcs. Proc. Amer. Math. Soc. 21, 310–314 (1969)
16. Lutz, F., Ziegler, G.: A small Polyhedral $\mathbb{Z}$-Acyclic 2-complex in $\mathbb{R}^4$. EG-Models, No. 2008.11.001 (2008)
17. Molina-Abril, H., Real, P.: Homological Spanning Forest Framework for 2D Image Analysis. To Appear in Ann. Math. Artif. Intell.
18. Mrozek, M., Pilarczyk, P., Żelazna, N.: Homology algorithm based on a cyclic subspace. Comput. Math. Appl. 55, 2395–2412 (2008)
19. Pacheco, A., Mari, J.L., Real, P.: Obtaining cell complexes associated to four dimensional digital objects. Imagen-A 1, 57–64 (2010)
20. Pitcher, E.: Inequalities of critical point theory. Bull. Amer. Math. Soc. 64, 1–30 (1958)
21. Robins, V., Wood, P.J., Sheppard, A.P.: Theory and algorithms for constructing discrete Morse complexes from grayscale digital images. IEEE Trans. Pattern Anal. Mach. Intell. 33(8), 1646–1658 (2011)
22. Whitehead, J.H.C.: Simple homotopy types. Amer. J. Math. 72, 1–57 (1950)

Removal Operations in nD Generalized Maps for Efficient Homology Computation

Guillaume Damiand[1], Rocio Gonzalez-Diaz[2], and Samuel Peltier[3]

[1] Université de Lyon, CNRS, LIRIS, UMR5205, F-69622, France
[2] Universidad de Sevilla, Dpto. de Matemática Aplicada I, S-41012, Spain
[3] Université de Poitiers, CNRS, XLIM-SIC, UMR6172, F-86962, France

Abstract. In this paper, we present an efficient way for computing homology generators of nD generalized maps. The algorithm proceeds in two steps: (1) cell removals reduces the number of cells while preserving homology; (2) homology generator computation is performed on the reduced object by reducing incidence matrices into their Smith-Agoston normal form. In this paper, we provide a definition of cells that can be removed while preserving homology. Some results on 2D and 3D homology generators computation are presented.

Keywords: nD Generalized Maps, Cellular Homology, Homology Generators, Removal Operations.

1 Introduction

In this paper, we propose a method for efficiently computing homology generators of subdivided cellular objects. The main idea is to simplify a subdivided object into a smaller one while preserving its homology. This principle is similar to the one used in [10] which is mainly algebraic (i.e. based on reduction of chain complexes), while our approach is mainly combinatorial.

In this work, we define a simplification algorithm based on the cell removal operations defined on generalized maps. Its principle is to simplify as much as possible the number of cells while preserving homology. Then we reduce incidence matrices (used for describing boundary operators) into their Smith-Agoston normal form for computing homology generators [3]. Moreover, generators computed in the reduced object can easily be projected into the original one.

The paper is structured as follows: in Sect. 2 all the necessary background regarding n-Gmaps is recalled. Section 3 presents the main result of the paper: the definition of the simplification algorithm based on the removal of two types of cells, and the proof of the homology preservation. Finally, some experiments are presented in Sect. 4 in order to illustrate that the simplification step widely reduces the number of cells, and also the homology generator computation.

2 Preliminary Works

An n-Gmap is a combinatorial structure devoted to the representation of cellular subdivision of orientable or not orientable nD quasi-manifolds, with or without

M. Ferri et al. (Eds.): CTIC 2012, LNCS 7309, pp. 20–29, 2012.

boundaries (see [11,12] for more details). Any polytopal complex can be described by an n-Gmap, while the converse is not true (an i-cell can be non homeomorphic to an i-disk). It is possible to associate a semi-simplicial set with any n-Gmap. An n-Gmap is not constructed directly from the cells of the subdivision but from more elementary objects: *darts*. The set of darts is structured through involutions that describe how they are linked to each other.

Definition 1 (n-Gmap). *An n-dimensional generalized map, called n-Gmap, with $0 \leq n$, is a $(n+2)$-tuple $G = (D, \alpha_0, \ldots, \alpha_n)$ where:*

1. *D is a finite set of darts;*
2. *$\forall i, 0 \leq i \leq n$, α_i is an involution on D;*
3. *$\forall i : 0 \leq i \leq n-2, \forall j : i+2 \leq j \leq n, \alpha_i \circ \alpha_j$ is an involution.*

The cells of the subdivision are defined implicitly as set of darts thank to the orbit notion (see Def. 2). An orbit in an n-Gmap can be seen as the set of darts that we can reach from a given dart and using as many times as possible the given involutions.

Definition 2 (Orbit). *Let $\Phi = \{\pi_0, \cdots, \pi_n\}$ be a set of permutations defined on a set D. $\langle \Phi \rangle$ is the permutation group of D generated by Φ. The orbit of an element $d \in D$ relatively to $\langle \Phi \rangle$, denoted $\langle \Phi \rangle (d)$ is the set $\{\phi(d) \mid \phi \in \langle \Phi \rangle\}$.*

As we can see in Def. 3, each i-dimensional cell is an n-Gmap is obtained by an orbit using all the involutions except α_i.

Definition 3 (i-cell). *Let G be an n-Gmap, and $d \in D$ be a dart. Given i, $0 \leq i \leq n$, the i-dimensional cell containing d, called i-cell and denoted by $c^i(d)$, is $\langle \alpha_0, \ldots, \alpha_{(i-1)}, \alpha_{(i+1)}, \ldots, \alpha_n \rangle (d)$.*

Due to the definition of cells as sets of darts, the incident and adjacency relations on cells can easily be tested. Two distinct cells c_1 and c_2 are *incident* if $c_1 \cap c_2 \neq \emptyset$, and two i-cells c_1 and c_2 are *adjacent* if there is two darts $d_1 \in c_1$ and $d_2 \in c_2$ satisfying $d_1 = \alpha_i(d_2)$. When a dart d belongs to an i-dimensional border, we have $\alpha_i(d) = d$ and we say that d is i-free.

In the example of Fig. 1, face f_3 is described by $\langle \alpha_0, \alpha_1 \rangle (1) = \{1, 2, 3, 4, 5, 6\}$, edge e_1 by $\langle \alpha_0, \alpha_2 \rangle (13) = \{13, 14, 15, 16\}$, and vertex v_1 by $\langle \alpha_1, \alpha_2 \rangle (2) = \{2, 3, 7, 14, 15, 24\}$. v_1 and e_1 are incident since $\langle \alpha_1, \alpha_2 \rangle (2) \cap \langle \alpha_0, \alpha_2 \rangle (13) = \{14, 15\} \neq \emptyset$. f_1 and f_3 are adjacent since $23 \in f_1$, $1 \in f_3$, and $\alpha_2(1) = 23$.

In this paper, the main operations used to simplify an n-Gmap are the *removal operations* (see [7,6] for the definitions). Intuitively, removing a removable cell c merges the two $(i+1)$-cells incident to c, without modifying the other cells.

Definition 4 (Removable cell). *Let G be an n-Gmap, c be an i-cell of G. c is removable if one of the two conditions is satisfied:*
$$i = n - 1; \text{ or } 0 \leq i < n - 1 \text{ and } \forall d \in c, \alpha_{i+1} \circ \alpha_{i+2}(d) = \alpha_{i+2} \circ \alpha_{i+1}(d).$$

The notion of removable cell c is strongly related to the number of its $(i + 1)$ incident cells, called the *degree* of c and denoted $degree(c)$. A direct consequence of Def. 4 is that an i-cell c of degree > 2 is not removable.

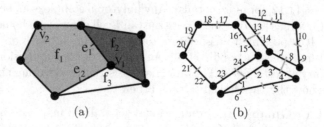

Fig. 1. Example of a 2G-map $G = (D, \alpha_0, \alpha_1, \alpha_2)$. (a) A 2D cellular complex containing 3 faces; 9 edges and 7 vertices. (b) The 2G-map describing this cellular complex, having 24 darts (represented by numbered black segments). Two darts linked by α_0 are drawn consecutively and separated by a gray segment (for example $\alpha_0(19) = 20$), two darts linked by α_1 share a common point (for example $\alpha_1(20) = 21$), and two darts linked by α_2 are drawn parallel, the gray segment over these two darts (for example $\alpha_2(13) = 16$).

In the example of Fig. 1, all the edges are removable (since an $(n - 1)$-cell is always removable in an nG-map), vertex v_2 is removable while vertex v_1 not. Removing edge e_1 merges faces f_1 and f_2 in one face having as boundary the boundary of f_1 plus the boundary of f_2 minus edge e_1.

To be able to compute homology of an n-Gmap, we need to have a boundary operator (defined in [5,4]). The boundary operator is defined for n-Gmaps having orientable cells. Note that it is possible to represent a non-orientable object (e.g. a Klein bottle) with a n-Gmaphaving only orientable cells.

In the following we detail the notions of orientable cell and signed cell (cf. Defs. 5 and 6).

Definition 5 (Orientable i-cell). *An i-cell c is* orientable *if $c = e_1 \cup e_2$ such that: $\forall d \in c$, $\forall j$, $0 \leq j \leq n$, $j \neq i$: d is not j-free $\Rightarrow$ d and $\alpha_j(d)$ do not belong to the same set e_1 or e_2. c is* non-orientable *otherwise.*

If c is orientable, then it can be partitioned in two sets of darts representing its two orientations and we can associate a value -1 or $+1$ to each of its dart, called a *sign*. In the following, we only consider n-Gmap having all its cells signed.

Definition 6 (Signed i-cell). *Let c be an orientable i-cell. The corresponding signed i-cell is c together with a sign for each of its dart d, denoted $sg^i(d)$:*

- *$sg^i(d) = -sg^i(\alpha_j(d))$ $\forall j: 0 \leq j < i$ such that d is not j-free;*
- *$sg^i(d) = sg^i(\alpha_j(d))$ $\forall j: i < j \leq n$.*

For defining a boundary operator on n-Gmaps, we first define the signed incidence number between two cells c^i and c^{i-1} which describes the number of times that c^{i-1} appears in the boundary of c^i.

Definition 7 (Signed incidence number). *let $\{p_j\}_{j=1\cdots k}$ be a set of darts s.t. the orbits $\{\langle \alpha_0, \cdots, \alpha_{(i-2)} \rangle (p_j)\}_{j=1\cdots k}$ make a partition of $\langle \alpha_0, \ldots, \alpha_{(i-1)} \rangle (d)$. The* signed incidence number *between c^i and c^{i-1} is defined by*

$$(c^i : c^{i-1}) = \sum_{p_j, j=1\cdots k | p_j \in c^{i-1}} sg^i(p_j).sg^{i-1}(p_j).$$

Note that this definition is equivalent to the one given in [5]. Now the boundary operator ∂_G of any i-cell c is defined as $\partial_G(c) = \sum_{c'}(c : c')c'$, where c' are $(i-1)$-cells incident to c. The boundary operator ∂_G satisfies $\partial_G \circ \partial_G = 0$ when involutions α_i are without fixed points for $0 \le i \le n-1$. Moreover, we have proven in [4] that the homology defined on n-Gmaps by this boundary operator is equivalent to the simplicial homology of the associated quasi-manifolds when the homology of the canonical boundary of each i-cell is that of an $(i-1)$-sphere, and when $\forall d \in D$, $\forall i \in \{0, \ldots, n\}$, d is i-free or $\alpha_i(d) \notin \langle \alpha_0, \ldots, \alpha_{i-2}, \alpha_{i+2}, \ldots, \alpha_n \rangle (d)$. In the following, all the considered n-Gmaps satisfied these conditions.

3 Removal Operations Preserving Homology

In this section, we prove that removing a degree two cell or a dangling cell preserves the homology of the n-Gmap.

3.1 Chain Complexes and Chain Contractions

Let $S = \{S_q\}_q$ be a graded. A q-chain is a finite formal sum of elements of S_q with coefficients in $\mathbb{Z}$. Let $C_q(S)$ denote the group of q-chains of S. The *chain complex* $(C_*(S), \partial)$ is the chain group $C_*(S) = \{C_q(S)\}_q$ together with a boundary operator ∂. Given an n-Gmap G, let S_G be the set of all the cells of G. $(C_*(S_G), \partial_G)$ is the chain complex associated to G.

A *chain contraction* [13] of $(C_*(S), \partial)$ to $(C_*(S'), \partial')$ is a triple $(f = \{f_q : C_q(S) \to C_q(S')\}_q$, $g = \{g_q : C_q(S') \to C_q(S)\}_q$ and $\phi = \{\phi_q : C_q(S) \to C_{q+1}(S)\}_q)$ such that: (i) f and g are chain maps; i.e. $f_q \circ \partial_q = \partial'_q \circ f_q$ and $g_q \circ \partial_q = \partial'_q \circ g_q$ for all q; (ii) ϕ is a chain homotopy of $id_{C_*(S)} = \{id_q : C_q(S) \to C_q(S)\}_q$ to $g \circ f = \{g_q \circ f_q : C_q(S) \to C_q(S)\}_q$; i.e. $\phi_{q-1} \circ \partial_q + \partial'_{q+1} \circ \phi_q = id_q - g_q \circ f_q$ for all q; (iii) $f \circ g = id_{C_*(S')}$. If a chain contraction of $(C_*(S_G), \partial_G)$ to $(C_*(S_{G'}), \partial_{G'})$ exists, then the n-Gmaps G and G' have isomorphic homology groups.

3.2 Degree Two Cells

Proposition 1. *Let c be an i-cell in an n-Gmap. If c is removable and degree two cell, then there are two $(i+1)$-cells a and b satisfying: $|(a : c)| = |(b : c)| = 1$ and for all other $(i+1)$-cells c', $(c' : c) = 0$.*

Proof. Since c is degree two, there are two $(i+1)$-cells a and b that are incident to c. For these two cells, we have $c \in \partial_G(a)$ and $c \in \partial_G(b)$. So, $(a : c) \ne 0$ and $(b : c) \ne 0$. If $|(a : c)| > 1$, contradiction with removal property, thus $|(a : c)| = 1$ (and the same for $|(b : c)| = 1$). For all other $(i+1)$-cells c', c' is not incident to c otherwise the degree was greater than two. Thus $(c' : c) = 0$. □

Proposition 2. *Let c be an i-cell in an n-Gmap. If c is a removable degree two cell, and if each j-cell e incident to c, is after the removal of c a j-cell equal to $e \setminus c$, then homology is preserved after the removal of c.*

Note that the removal of a cell may induce removal of other cells (for example, it is possible to build a sphere made of one vertex, one degree two edge and two faces. Removing the edge would supress all the darts and so the vertex and the two faces). The second condition ensures that only one cell is removed

Proof. Let $(C_*(S_G), \partial_G)$ be the chain complex associated to G. Since c is degree two, there are two $(i+1)$-cells a and b that are incident to c. The set $S_{G'}$ of the cells of the n-Gmap G' obtained after removing the cell c consists in $S_G \setminus \{a, b, c\} \cup \{a'\}$ where a' is the resulting $(i+1)$-cell from merging the two cells a and b. Since, by Prop. 1, $|(a : c)| = |(b : c)| = 1$ and for all other $(i+1)$-cells c', $(c' : c) = 0$, we can construct a chain contraction (f, g, ϕ) of $(C_*(S_G), \partial_G)$ to $(C_*(S_{G'}), \partial_{G'})$ as follows:

$$f(x) = \begin{cases} c - (b:c)\partial_G(b), & \text{if } x = c, \\ a', & \text{if } x = a, \\ 0, & \text{if } x = b, \\ x, & \text{otherwise;} \end{cases}$$

$$g(x) = \begin{cases} a - (a:c)(b:c)b, & \text{if } x = a', \\ x, & \text{otherwise;} \end{cases}$$

$$\phi(x) = \begin{cases} (b:c)b, & \text{if } x = c, \\ 0, & \text{otherwise.} \end{cases}$$

To check that (f, g, ϕ) is a chain contraction is left to the reader. Moreover, we know that each j-cell incident to c is preserved by the removal operation. Then G and G' have isomorphic homology groups.   □

3.3   Dangling Cells

Let $(C_*(S), \partial)$ be a chain complex. Let $s, t \in S$ such that $|(s : t)| = 1$ and $(s' : t) = 0$ for any $s' \in S$, $s' \neq s$. If we remove s and t from S to get S', we obtain another chain complex $(C_*(S'), \partial')$ which is called an *elementary collapse* of S. A chain contraction of $(C_*(S), \partial)$ to $(C_*(S'), \partial')$ is given by

$$f(x) = \begin{cases} 0, & \text{if } x = s, \\ t - (s:t)\partial(s), & \text{if } x = t, \\ x, & \text{otherwise;} \end{cases} \quad g(x) = x; \quad \phi(x) = \begin{cases} (s:t)t, & \text{if } x = t, \\ 0, & \text{otherwise.} \end{cases}$$

Therefore an elementary collapse preserves homology. A subset of S is *collapsible* if they can all be removed from S in a sequence of elementary collapses.

Let c be a k-cell, the *closure* of c, denoted $\bar{c}$, is the set made of c plus all the j-cells, $0 \leq j < k$ that are incident to c. The closure of a set S of cells, denoted $\bar{S}$, is the union of the closures of all the cells of S.

Definition 8 (Dangling cell). *Let c be an i-cell. We denote C the set of $(i-1)$-cells of $\partial_G(c)$, and $B = \{c' \in \partial_G(c) | degree(c') > 1\}$. c is dangling if c is orientable, its degree is 1, $\{c\} \cup \overline{C} \setminus \overline{B}$ is collapsible, and each j-cell $e \in \overline{B}$, is after the removal of c a j-cell equal to $e \setminus c$.*

Proposition 3. *Let c be an i-cell in an n-Gmap. If c is removable and dangling cell, then its removal preserves the homology of the n-Gmap.*

Proof. Removing c will remove also all the cells in $\overline{C} \setminus \overline{B}$ because these cells are included in c (i.e. their set of darts is included in the set of darts of c). As $\{c\} \cup \overline{C} \setminus \overline{B}$ is collapsible, and as all the other cells are preserved, the homology of the n-Gmap is preserved by the definition and property of collapsible.   □

3.4   Simplification Preserving Homology

The main principle of the simplification algorithm consists in removing successively all the degree two cells and all the dangling cells for all the dimensions starting from $(n-1)$-cells to 0-cells. For that, we start to define Algo. 1 which simplifies all the i-cells of a given n-Gmap for a given dimension i.

Algorithm 1. Simplification of i-cells.

Input: An n-Gmap G.
Output: Simplify all the i-cells of G while preserving the same homology.

foreach *i-cell c of* G **do**
 if *c is removable* **and** *the degree of c is 2* **then**
  Remove f;
 else if *c is removable* **and** *c is a dangling cell* **then**
  push(P, c);
  repeat
   $c \leftarrow$ pop(P);
   push in P all the dangling i-cells adjacent to c;
   Remove c;
  until *empty(P)*;

In this algorithm, we consider successively each i-cell c, and there are three possible cases. First, if c is not removable, then we are sure that c cannot be removable in a future step of the algorithm. Indeed, we only remove i-cells and this does not modify the $(i+1)$-cells incident to c. Second, if c is removable and its degree is two, we remove c. Third, if c is removable and dangling, we also remove c, but now we have to reconsider all the i-cells adjacent to c. Indeed, these cells can possibly become dangling due to the removal of c. At the end of the loop, we have considered all the i-cells and removed all the degree 2 cells and the dangling cells that were removable.

Now the global simplification method consists only in simplifying all the i-cells of the n-Gmap for all the cells by decreasing dimensions. We have to work in decreasing dimensions because the removal of an i-cell modifies the degree of all the incident $(i-1)$-cells. At the end of the global simplification algorithm, we have removed all the removable cells of degree 2 or dangling. By using Props. 2 and 3, we know that the final n-Gmap obtained after all the removals has the same homology than the initial n-Gmap.

4   Experiments

In order to illustrate the interest of our simplification algorithm, we show results on homology generator computation for the five objects shown in Fig. 2. Objects (a), (b) and (c) are described by 2-Gmaps; objects (d) and (e) are described by 3-Gmaps.

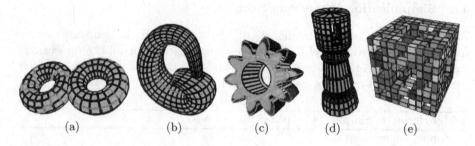

Fig. 2. (a) 2-torus. (b) Klein bottle. (c) pinion. (d) tower. (e) Menger sponge.

Table 1. Results of our experiments. We give the number of cells (columns # *cells*) for initial objects, and after the simplification algorithm. The last column gives the time of the simplification step. The two columns *Homology computation* give the memory space and the time of the homology generators computation (0s means less than 10^{-6}s).

Object	Initial						Simplified					
	# cells				Homology		# cells				Homology	Simplif.
Cell dim.	0	1	2	3	computation		0	1	2	3	computation	time
2-torus	404	802	396	-	14Mb	5.76s	6	9	1	-	2.36Kb 0s	0s
Klein	900	1800	900	-	74Mb	128.47s	2	3	1	-	0.41Kb 0s	0s
Pinion	470	701	231	-	11Mb	3.56s	2	3	1	-	0.41Kb 0s	0s
Tower	906	1856	952	4	85Mb	140.97s	10	15	4	1	6.53Kb 0s	0s
Menger	896	2304	1728	400	159Mb	372.50s	189	365	97	1	2938.00Kb 0.81s	0.03s

To compute the homology generators, we iterate through all the cells of the n-Gmap and we compute incidence matrices (which describes the boundary of the cells) using the incidence number definition. Then we reduce incidence matrices into their Smith-Agoston normal form for computing homology generators [3]. Compared to the classical Smith normal form, the specificity of the Agoston reduced normal form is that for a given dimension d, the basis of the boundaries B_p is a subset of the basis of cycles Z_p, thus the quotient group $H_p = Z_p/B_p$ can directly be obtained by simply removing from Z_p the boundaries of infinite order. Note that several optimizations exists for the reduction of incidence matrices [15,8]. Even if they can be used, we do not use them here as we focus on showing the improvement obtained with the simplification process.

The computation of homology generators was implemented in Moka [16], a 3D topological modeler based on 3-Gmap. For this reason, the computation of homology generators is limited to 2D and 3D cases, but all the functions are generic in any dimension. The results are presented in Table 1, where the simplification step widely reduces the number of cells. On the last column one can see that the simplification step is very fast. Memory space is also reduced as the size of incidence matrices are directly linked to the number of cells.

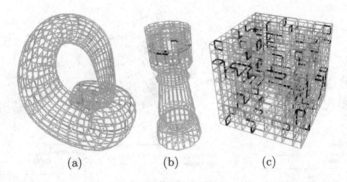

Fig. 3. The generators of H_1 (in red) computed on simplified objects, and projected on initial objects (drawn in grey). (a) Klein bottle. (b) Tower. (c) Menger sponge.

Lastly, we can see in Fig. 3 the different generators of H_1 obtained for some objects. By using the definition of removal operations, we are able to project the generators of the simplified object on the initial one (by using a similar technique as in [14,9]).

We have made a second type of experiments in order to compare our approach with other existing methods. To our knowledge there is no other general method which compute homology generator of cellular objects. Thus we compare our solution with Chomp and RedHom [1,2] which compute homology generators of cubical complexes. We chose these two methods since the two softwares are publicly available. However, it must be noticed that representing a cubical complex by a n-Gmaps is not efficient since a cube is described by 48 darts; the interest of cellular model is precisely to represent non regular subdivisions. In order to test the scale up property of the three methods, we chose three objects (see in Fig. 4(a), (b) and (c)), and multiply the size of each voxel by 4 to 9 for the first two objects, and by 2 to 7 for the last object which contains more voxels.

We can see in Fig. 4 the time required to compute homology generators of each object by the three compared methods. These results are really encouraging for our method which obtain the best computation time for the first two objects, with an important gain for the second one. For the third object, Chomp, and Moka performances are very similar (even if Chomp is a little bit quicker), while RedHom is really faster. In this last case, Chomp, and Moka have similar computation time than for the two first objects, while RedHom is extremely fast. We suppose there is an optimization allowing to remove directly some block of voxels. Indeed, the upper part of the last object is composed by a full block of voxels. This kind of improvement can also be made for our method.

These experiments show that our method is very competitive since it is not optimized for a specific type of subdivision but it is generic for any cellular complex. Thus its main interest is its genericity and we can conclude from this comparison that this is not to the detriment of the efficiency. Moreover, we can improve our results by adding a thinning pre-processing step that reduces the number of voxels while preserving the homology.

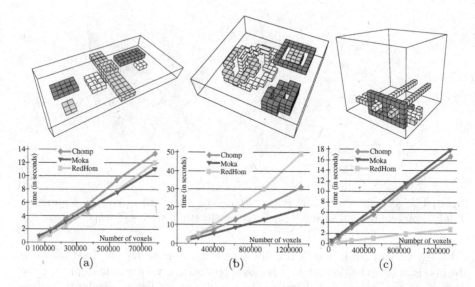

Fig. 4. Homology computation time comparison for `Chomp`, `RedHom` and `Moka`. Objects are made of voxels filling the bounding box (in wireframe), the filled surfaces being borders of cavities or tunnels. (a) `Cub1`: 1067 voxels; 1 connected components; 9 tunnels; 5 cavities. (b) `Cub2`: 1828 voxels; 1 connected components; 7 tunnels; 4 cavities. (c) `Cub3`: 4003 voxels; 1 connected components; 6 tunnels; 3 cavities.

5   Conclusion

In this paper, we have presented an algorithm that simplifies an n-Gmap while preserving its homology. For that, it removes degree two cells and dangling cells. Then we can compute homology on the reduced n-Gmap and project the generator on the original object. Some results show the interest of the simplification step, both in memory space and in computation time.

Some questions are still open. The first question is about the conditions on removed cells. Is it possible to remove some other type of cells while preserving the homology? The answer is no in 2D and 3D, but still open in higher dimension. This question is related to the definition of the minimal generalized map having the same homology. In 3D, to obtain this minimal map, we need to use another type of operation (fictive edge shifting). Thus we would like to study the extension of this operation in higher dimension to define the minimal n-Gmap.

References

1. Chomp, http://chomp.rutgers.edu/
2. Redhom, http://redhom.ii.uj.edu.pl/
3. Agoston, M.K.: Algebraic Topology, a first course. In: Dekker, M. (ed.) Pure and Applied Mathematics (1976)

4. Alayrangues, S., Damiand, G., Lienhardt, P., Peltier, S.: A boundary operator for computing the homology of cellular structures. Discrete & Computational Geometry (under submission)
5. Alayrangues, S., Peltier, S., Damiand, G., Lienhardt, P.: Border Operator for Generalized Maps. In: Brlek, S., Reutenauer, C., Provençal, X. (eds.) DGCI 2009. LNCS, vol. 5810, pp. 300–312. Springer, Heidelberg (2009)
6. Damiand, G., Dexet-Guiard, M., Lienhardt, P., Andres, E.: Removal and contraction operations to define combinatorial pyramids: Application to the design of a spatial modeler. Image and Vision Computing 23(2), 259–269 (2005)
7. Damiand, G., Lienhardt, P.: Removal and Contraction for n-Dimensional Generalized Maps. In: Nyström, I., Sanniti di Baja, G., Svensson, S. (eds.) DGCI 2003. LNCS, vol. 2886, pp. 408–419. Springer, Heidelberg (2003)
8. Dumas, J.-G., Heckenbach, F., Saunders, B.D., Welker, V.: Computing simplicial homology based on efficient smith normal form algorithms. In: Algebra, Geometry, and Software Systems, pp. 177–206 (2003)
9. Gonzalez-Diaz, R., Ion, A., Iglesias-Ham, M., Kropatsch, W.G.: Invariant representative cocycles of cohomology generators using irregular graph pyramids. Computer Vision and Image Understanding 115(7), 1011–1022 (2011)
10. Kaczynski, T., Mrozek, M., Slusarek, M.: Homology computation by reduction of chain complexes. Computers & Math. Appl. 34(4), 59–70 (1998)
11. Lienhardt, P.: Topological models for boundary representation: a comparison with n-dimensional generalized maps. CAD 23(1), 59–82 (1991)
12. Lienhardt, P.: N-dimensional generalized combinatorial maps and cellular quasi-manifolds. Computational Geometry & Applications 4(3), 275–324 (1994)
13. MacLane, S.: Homology. Classic in Mathematics. Springer (1995)
14. Peltier, S., Ion, A., Kropatsch, W.g., Damiand, G., Haxhimusa, Y.: Directly computing the generators of image homology using graph pyramids. Image and Vision Computing 27(7), 846–853 (2009)
15. Storjohann, A.: Near optimal algorithms for computing smith normal forms of integer matrices. In: Lakshman, Y.N. (ed.) Proceedings of the 1996 Int. Symp. on Symbolic and Algebraic Computation, pp. 267–274. ACM (1996)
16. Vidil, F., Damiand, G.: Moka (2003), http://moka-modeller.sourceforge.net/

Enhancing the Reconstruction from Non-uniform Point Sets Using Persistence Information

Erald Vuçini

VRVis Center for Virtual Reality and Visualization Research
vucini@vrvis.at

Abstract. In this paper we propose an efficient method for selecting the reconstruction resolution of non-uniform representations. We analyze the topological difference between reconstructions based on Topological Persistence information and define a distance for quantifying such information. We compute the Persistence information with a state-of-the-art method and report in detail the characteristics of the proposed algorithm. We evaluate our method in different scenarios and compare to previous contributions. Our proposed method offers faster and more reliable results in an effort to improve the reconstruction process and to reduce the necessity for visual inspection.

Keywords: Reconstruction, Topology, Persistence, Bottleneck Distance.

1 Introduction

Nowadays, the number of applications that provide non-uniform (irregular) data is increasing steadily. Examples vary from astronomical, Doppler and ultrasound measurements, to particle or numerical simulations in physical sciences. Non-uniform data representations offer a way of adapting the measure location according to the importance of the data. However, most of the techniques that deal with the analysis and processing of data are fitted to uniform (regular) data. The transform of non-uniform representations to uniform ones, is a viable option, for the better understanding and analysis of such data. In this paper we focus on the process of reconstruction from non-uniform to uniform representations.

The main problem in the reconstruction from non-uniform representations is the selection of the proper resolution of reconstruction. This will be the central question we will try to answer in this paper. The problem of selection of reconstruction resolution can be translated into a trade-off finding problem between accuracy and memory efficiency. A coarse (low) resolution of reconstruction requires less memory consumption, but will smooth the signal in areas with sharp transitions, resulting in visual artifacts or high errors. On the other side, a too fine (high) resolution of reconstruction will introduce memory overheads while achieving a better accuracy.

Another important issue, in the reconstruction process is the evaluation of the quality of reconstruction. A commonly followed approach for assessing the quality of reconstruction is to measure the reconstruction error, e.g., the root mean square error (RMSE). In addition to the fact that the RMSE can be misguiding due to his averaging behavior, in many scenarios low RMSEs may still result in artifacts in the reconstructed data. Hence, direct visual inspection is required, introducing the need of more time

M. Ferri et al. (Eds.): CTIC 2012, LNCS 7309, pp. 30–38, 2012.

overhead and user interaction. In a traditional reconstruction framework, we would follow these steps when trying to reconstruct a non-uniform points set **P**: (1) reconstruct **P** with different resolutions, (2) compute the RMSEs for each reconstruction resolution, (3) select those reconstructions that have RMSEs lower than a user defined threshold, and (4) visually inspect the reconstructed data for possible artifacts, e.g., by means of volume rendering. Vuçini and Kropatsch [17] proposed to reduce the necessity for visual inspection by using topological information derived from Homology analysis. A schematic view of this reconstruction pipeline is displayed in Fig. 1.

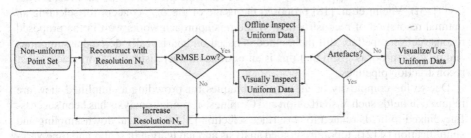

Fig. 1. Schematic view of a pipeline for the reconstruction of non-uniform point sets to uniform representations when the target resolution is unknown. The offline inspection step represents the option to decide over the quality of reconstructions using additional information and without the need of visual inspection.

In this paper, we build upon the work of Vuçini and Kropatsch [17]. We also derive a topological signature from the analysis of data. Instead of homology we use persistent homology (or simple Persistence) in our analysis workflow. The usage of Persistence allows us to compute a more robust topological signature, in a more efficient way due to recent developments in persistence computation algorithms [20]. The topological information together with error measurements improves the quality assessment of the reconstruction and reduces the need of visual inspection.

In Section 2 we give a short summary of related work. We introduce our topology controller in Section 3, and explain the main modules of the proposed algorithm. In Section 4 we show results w.r.t. the usage of the topological signature and assess the value of its applicability. Finally, conclusions are drawn and some ideas are layed out and discussed as future work in Section 5.

2   Related Work

Following we give an overview of work related to the proposed approach, namely, with regard to reconstruction and topology analysis of data.

Non-uniform data reconstruction (approximation) is a recent, fast growing research area. A number of approaches reconstruct non-uniformly sampled data, especially for one- and two-dimensional signals. Most of the methods are based on the reconstruction of the data by solving large systems of equations ([11]). Perhaps the most popular approach for approximating non-uniform data is based on Radial Basis Functions (RBFs). They have been used in surface ([14]) as well as volumetric ([13]) approximation and reconstruction techniques. Arigovindan et al. [2] proposed to use B-splines in a multi-grid framework for the reconstruction of non-uniform 2D data. Vuçini et al. [18] extended these ideas to 3D volumes and large datasets.

Most of the above-mentioned approaches consider the resolution of reconstruction as known *a priori*. In order to find the resolution allowing exact reconstruction a lower bound on the minimal distance between two sampling positions has to be assured. For general shift-invariant spaces a Beurling density $D \geq 1$ is necessary for a stable and perfect reconstruction ([1]). In topology analysis, in order to be able to provide a topological-stable reconstruction, the object (signal) taken into consideration has to be r-regular. The related literature is mainly related to the problem of surface reconstruction ([15]). Vuçini et al. [18] proposed the usage of the σ_{avg} concept for selecting an optimal resolution of reconstruction. While this approach works well in the proposed reconstruction pipeline, still the method is based on heuristically derived assumptions and no clear proof is given that this is an optimal characteristic that works with other reconstruction pipelines.

Due to the complexity of the data, techniques for providing a simplified view are required in fields such Visualization and Graphics. Topology analysis has been successfully linked to fields related to isosurface selection [3], topological downsampling and simplification ([12]), topology-guided analysis and navigation in scalar and time varying data ([5]), and feature tracking and evolution ([21]). Carr et al. [6] have presented a generalized framework consolidating the theory and application of the contour spectrum concept. Most of the above-mentioned works have concentrated in reporting topological information related to 0-dimensional homology, i.e., connected components. Topological persistence information has been also used for shape comparison and feature classification ([7], [8], [4]). In our previous work ([17]), homology information was used to derive a topological signature. The main drawback of the method, is the necessity to compute the homology information for each superlevel set in order to derive important information. In the current approach the usage of Persistence overpasses this drawback, resulting in lower computation times and in the same time providing a more compact representation.

All the above-mentioned methods provide extensive information, which is difficult to interpret without the appropriate statistical analysis. Similar to [17], in the proposed method we use statistical topological information for guiding the selection of resolution of reconstruction. As a result, our framework gives important cues that reduce the necessity of human's visual inspection of the data.

3   Topology-Based Analysis

Our proposed algorithm consists of two main modules: 1) the variational reconstruction module, and 2) the module that derives the statistical Persistence-based information. Both modules are integrated in the main iterative procedure which extracts useful statistical information related to the reconstruction process. Through this information we will be able to select a resolution of reconstruction that has both a low RMSE and topological stability with regards to our defined Persistence-based topology-controller ($\mathscr{PC}$).

3.1   Variational Reconstruction Basics

Variational reconstruction is a well-known technique applied to solving ill-posed problems such as the reconstruction from non-uniform point sets. The variational functional is formulated so that it provides a solution close to the input points, while regularizing the smoothness in order to prevent discontinuities.

Given a set $\mathbf{S}$ of sample points, $\mathbf{s}_i = (x_i, y_i, z_i)$, $i = 1, 2, \ldots, M$, the B-spline approximation is formulated in a way that it approximates the points set in a resolution (N_x, N_y, N_z) of the axis-aligned bounding box. Cubic B-splines do not enjoy the interpolation property, but with real-world data where noise is always present, approximative (not-interpolating) splines are better suited for the reconstruction process ([16]).

The key idea of the variational reconstruction is to build a linear system and to solve it by minimizing a derived cost function. Once the linear system is solved, we can estimate the approximating function $f(\mathbf{s})$ (a C^2-continuous function) at any position $\mathbf{s} \in V$, where V is the volume enclosing the bounding box of the non-uniform point set. For a deeper insight into the method we refer the reader to [18].

3.2 Persistence-Based Topology Analysis

We shortly introduce Persistence, focusing on $\mathbb{Z}_2$ homology ([10]). Given a topological space $\mathbb{X}$ and a *filtering function* $f : \mathbb{X} \to \mathbb{R}$, Persistence performs a topological exploration along a filtration, i.e., a nested sequence of subsets $X_1 \subseteq X_2 \subseteq \ldots \subseteq X_n = X$, usually induced by considering the sublevel sets of the filtering function. The algorithm captures the birth and death times of homology classes of the sublevel set as it grows along the filtration. By birth, we mean that a homology class comes into being; by death, we mean it either becomes trivial or becomes identical to some other class born earlier. The persistence, or lifetime of a class, is the difference between the death and birth times. Homology classes with larger persistence reveal information about the global structure of the space $\mathbb{X}$, as described by the function f.

3.3 Persistence-Based Algorithm (PbA)

Similar to [17], the proposed algorithm takes as input a non-uniform point set ($\mathbf{S}$), a minimum (N_{min}) and maximum (N_{max}) resolution of reconstruction, and a resolution step Δ (see Algorithm 1). In difference from [17] we do not require the number of superlevel sets needed to reconstruct. The number of superlevel sets was previously required by the Homology module, due to the fact that homology computation runs only on binary data, e.g., object and background. Persistence, offers us the possibility to create a filtration along the values of the data.

Algorithm 1. PbA($\mathbf{S}, N_{min}, N_{max}, \Delta$)

1: **for** $i = N_{min}$ to N_{max} **do**
2:     determine V_i by solving variational reconstruction for $\mathbf{S}$ on resolution i
3:     compute Persistence $\mathscr{P}^0_{V_i}$, $\mathscr{P}^1_{V_i}$ and $\mathscr{P}^2_{V_i}$
4:     build Persistence frequency histograms $\mathbf{PH}^0_{V_i}$, $\mathbf{PH}^1_{V_i}$ and $\mathbf{PH}^2_{V_i}$
5:     $i = i + \Delta$
6: **end for**
7: compute the topology controller ($\mathscr{PC}$)

The iterative algorithm starts with determining the volume V_i, from the approximating function estimated from the variational reconstruction of the non-uniform point set **S** (line 2). The resolution of reconstruction $N_x \times N_y \times N_z$ is specified by the loop-variable i (loop-variable i is augmented in each step by the variable Δ). By varying N_x, N_y and N_z are determined automatically by the proper aspect ratio of the axis-aligned bounding box enclosing the given non-uniform data points. In (line 3) we compute the persistence of the reconstructed volumes. As a result, we obtain a set of pairs for each dimension, i.e., $\mathcal{P}_{V_i}^0$, $\mathcal{P}_{V_i}^1$ and $\mathcal{P}_{V_i}^2$. Each pair represents a persistent topological feature and is given as two functional values, i.e., (birth, death). Following, we can build the Persistence frequency histograms $\mathbf{PH}_{V_i}^0$, $\mathbf{PH}_{V_i}^1$ and $\mathbf{PH}_{V_i}^2$ (line 4). The frequency histograms span the data range, e.g., in many data examples [0,4095], and measure the number of persistent topological features that are alive at a specific functional value. $\mathbf{PH}_{V_i}^\tau$ can be considered as the τ-dimensional persistence signature of the point set **S** in resolution i.

After computing the Persistence frequency histograms for each resolution, we can define the topology controller $\mathcal{PC}(i)$ as:

$$\mathcal{PC}(i) = \frac{1}{\alpha_0 + \alpha_1 + \alpha_2} \sum_{\tau=0}^{2} \alpha_\tau \cdot \frac{\left\| \mathbf{PH}_{V_i}^\tau - \mathbf{PH}_{V_{N_{max}}}^\tau \right\|}{\left\| \mathbf{PH}_{V_{N_{max}}}^\tau \right\|} \qquad (1)$$

where the weights (coefficients) $\alpha \in \{0,1\}$ control the impact of the respective τ-dimensional persistence statistics ($\mathbf{PH}^\tau$) on the topology controller. In simpler words, $\mathcal{PC}(i)$ computes the relative error of $\mathbf{PH}_{V_i}^\tau$ with regard to $\mathbf{PH}_{V_{N_{max}}}^\tau$, which is the persistence signature of the point set **S** in the maximum resolution.

4   Implementation and Results

Our test platform is an Intel i7 CPU @ 2.67GHz with 12GB of RAM. All the algorithms are developed as single threaded hence only one processor core is used. Analog to [17], we tested our framework with 3D data sets based on non-uniform point sets as well as Cartesian grids. For detailed information on the datasets refer to [19].

In our framework, we analyze graphical plots of $\mathcal{PC}$ with regard to a changing resolution and we set default weight values ($\alpha_0 = \alpha_1 = \alpha_2 = 1$) in Eq. 1. We attach to these plots also the graphs of $\mathcal{PC}_\tau$, which measure the τ-dimensional homological statistics. $\mathcal{PC}_\tau$ is derived from $\mathcal{PC}$ by setting the respective α_τ equal to one and the other two weights equal to zero. Vuçini and Kropatsch [17], suggested a threshold of 0.2 for the topology controller, in order to find the optimal resolution of reconstruction. While this is a viable approach, in this work we followed a different strategy. We let the user define upon the plots of the topology controller, which the resolution that better suits her. In this way, we avoid hard thresholding, which is not optimal when dealing with different data types.

In Fig. 2 we show renderings of the point set and reconstructions from the Cooling Jacket dataset. The dataset was generated at AVL List GmbH in order to evaluate a cooling jacket design for a four cylinder diesel engine. This stationary flow simulation incorporates a heat transport solution in order to predict critical temperature regions within the engine. The original dataset is specified on an unstructured grid. The non-uniform point set consists of 1,537,898 points encoding the pressure of the flow data.

The topology of the data is very crucial for such an industrial simulation, and our algorithm should be able to distinguish between the cases displayed in Fig. 2 b-d). In Fig. 3(a), we display the plots of the topology controller derived from our proposed algorithm. We see that the (red) line representing $\mathscr{PC}$, stabilizes after the resolution $N_x = 410$. Hence, this is the value that we select as an optimal reconstruction resolution for the Cooling Jacket dataset. Visually comparing the renderings in Fig. 2 and 3(b), we can also confirm that the reconstruction done with the selected resolution is topologically more similar to the input point set, than the ones done with lower resolutions. In Table 1 we provide results of our framework for different datasets and also compare with the methods proposed in [18] and [17]. While our proposed resolutions are different from the ones suggested in [18], they are similar to the ones proposed in [17].

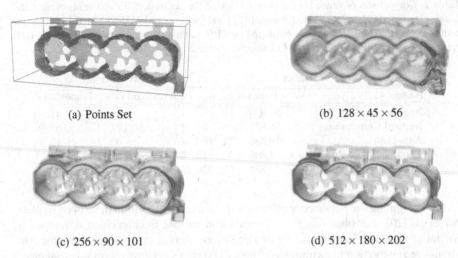

(a) Points Set   (b) $128 \times 45 \times 56$

(c) $256 \times 90 \times 101$ (d) $512 \times 180 \times 202$

Fig. 2. Cooling Jacket (Pressure) dataset: a) dataset consisting of 1,537,898 non-uniform points, b-d) reconstructions with resolution $128 \times 45 \times 51$, $256 \times 90 \times 101$ and $512 \times 180 \times 202$ respectively

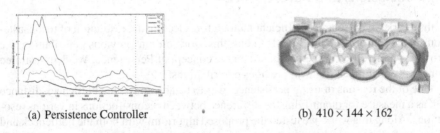

(a) Persistence Controller   (b) $410 \times 144 \times 162$

Fig. 3. Cooling Jacket (Pressure) dataset: a) Persistence-based controller, b) reconstruction with resolution $410 \times 144 \times 162$. The hair-line in the Fig. a) highlights the values of the Persistence Controllers for the selected resolution ($N_x = 410$).

Table 1. Comparisons of reconstruction resolutions for different non-uniform datasets as proposed by [18] and our proposed Topology controller

Dataset		Previous [18]		Previous [17]		Proposed	
Name	Points	Resolution	RMSE	Resolution	RMSE	Resolution	RMSE
Oil	29,094	38x40x38	0.19	41x44x41	0.16	43x45x43	0.15
Natural Convection	68,921	61x61x61	0.63	55x55x55	0.71	58x58x58	0.68
Synthetic Chirp	75,000	64x64x64	1.12	58x58x58	1.18	60x60x60	1.16
Bypass	7,929,856	766x92x192	0.61	880x106x220	0.49	860x103x216	0.55
Blunt-Fin	40960	93x36x25	1.14	102x39x28	1.11	98x38x26	1.12
Cooling Jacket	1,537,898	256x90x101	0.92	400x140x157	0.86	410x144x162	0.83

Table 2. Comparison of times for the computation of the controllers for different datasets and settings, as reported by Vuçini and Kropatsch[17] and the proposed method. (*) Timings for the results in the 'Previous' column are estimated on a PC, running on an Intel Dual Core 2.70 GHz processor machine with 6GB of RAM (similar single-core performance).

Dataset					Times (min)	
Name	Points	N_{min}	N_{max}	Δ	Previous [17](*)	Proposed
Neghip	52,428	16	80	1	60.45	1.47
Natural Convection	68,921	16	64	1	107.13	0.63
Aneurism	419,430	32	160	4	40.77	8.12
Cooling Jacket(Pressure)	1,537,898	32	512	2	7215.32	282.13
Bypass	7,929,856	256	1024	16	6712.80	257.98

For computing the Persistence information we use the algorithm developed by Wagner et al. [20]. For obtaining the reconstruction we use the algorithm developed by Vuçini et al. [18]. The complexity of Persistence algorithm is cubic, while the variational reconstruction algorithm has a linear complexity. However, in our framework we observed a linear complexity from both the modules. Computation times of the proposed framework, and comparison to [17] are given in Table 2. Note that, with the usage of the Persistence module we significantly decrease the computation times.

5   Conclusions and Future Work

In this work we presented an efficient method for selecting the resolution of reconstruction for non-uniform point sets. Building this work on our previous contribution, we provide a more efficient method based on the concept of Persistence. We demonstrated our results in comparison with previous contributions.

One of the reasons of using persistence, was to be able to use the bottleneck distance [9], as a measure for quantifying the difference between reconstructions in various resolutions. Along this work, we tested the proposed algorithm with both the bottleneck and Haussdorff distances. However, both preliminary testings resulted in a non-converging distance. In our future work, we want to analyze in more detail, the reasons underlying such fact. We believe, this can be related with the way our distance computing algorithm handles essential and non-essential classes in the persistence diagram. Another reason could be related to the possible normalizations, that have to be done in order to be able to compare different persistence diagrams.

In our future work, we also want to enable the usage of persistence-based topological signatures in visualization and graphics, e.g., in designing transfer functions or highlighting topological features of interest.

Aknowledgement. This work was supported by the Austrian Science Fund (FWF) grant no. P20134-N13 and the Austrian COMET program. The author thanks Dr. Andrea Cerri and Prof. Walter Kropatsch for the useful discussions related to the topic of this paper.

References

1. Aldroubi, A., Gröchenig, K.: Beurling-Landau-type theorems for non-uniform sampling in shift invariant spline spaces. Journal of Fourier Analysis and Applications 6, 93–103 (2000)
2. Arigovindan, M., Sühling, M., Hunziker, P.R., Unser, M.: Variational image reconstruction from arbitrarily spaced samples: A fast multiresolution spline solution. Proceedings of IEEE Transactions on Image Processing 14, 450–460 (2005)
3. Bajaj, C.L., Pascucci, V., Schikore, D.: The contour spectrum. In: Proceedings of IEEE Visualization, pp. 167–174 (1997)
4. Bendich, P., Edelsbrunner, H., Kerber, M.: Computing robustness and persistence for images. In: Proceedings of IEEE Visualization, pp. 1251–1260 (2010)
5. Bremer, P.-T., Weber, G.H., Pascucci, V., Day, M., Bell, J.B.: Analyzing and tracking burning structures in lean premixed hydrogen flames. IEEE Trans. Vis. Comput. Graph. 16(2), 248–260 (2010)
6. Carr, H., Snoeyink, J., van de Panne, M.: Flexible isosurfaces: Simplifying and displaying scalar topology using the contour tree. Computational Geometry 43(1), 42–58 (2010)
7. Cerri, A., Biasotti, S., Giorgi, D.: k-dimensional size functions for shape description and comparison. In: International Conference on Image Analysis and Processing, pp. 795–800 (2007)
8. Chazal, F., Cohen-Steiner, D., Guibas, L.J., Mémoli, F., Oudot, S.: Gromov-hausdorff stable signatures for shapes using persistence. Comput. Graph. Forum 28(5), 1393–1403 (2009)
9. Cohen-Steiner, D., Edelsbrunner, H., Harer, J.: Stability of persistence diagrams. Discrete and Computational Geometry 37(1), 103–120 (2007)
10. Edelsbrunner, H., Harer, J.: Computational topology, an introduction. American Mathematical Society (2010)
11. Feichtinger, H.G., Gröchenig, K., Strohmer, T.: Efficient numerical methods in non-uniform sampling theory. Numerische Mathematik 69, 423–440 (1995)
12. Gyulassy, A., Natarajan, V., Pascucci, V., Bremer, P.-T., Hamann, B.: Topology-based simplification for feature extraction from 3D scalar fields. In: Proceedings of IEEE Visualization, pp. 535–542 (2005)
13. Jang, Y., Botchen, R.P., Lauser, A., Ebert, D.S., Gaither, K.P., Ertl, T.: Enhancing the interactive visualization of procedurally encoded multifield data with ellipsoidal basis functions. Computer and Graphics Forum 25(3), 587–596 (2006)
14. Ohtake, Y., Belyaev, A.G., Seidel, H.-P.: 3D scattered data approximation with adaptive compactly supported radial basis functions. In: Proceedings of International Conference on Shape Modeling and Applications, pp. 31–39 (2004)
15. Stelldinger, P., Latecki, L.J., Siqueira, M.: Topological equivalence between a 3d object and the reconstruction of its digital image. IEEE Trans. Pattern Anal. Mach. Intell. 29(1), 126–140 (2007)

16. Thévenaz, P., Blu, T., Unser, M.: Interpolation revisited. IEEE Transactions on Medical Imaging 19(7), 739–758 (2000)
17. Vuçini, E., Kropatsch, W.G.: On the search of optimal reconstruction resolution, Pattern Recognition Letters (2011) (in press)
18. Vuçini, E., Möller, T., Eduard Gröller, M.: On visualization and reconstruction from non-uniform point sets using b-splines. In: Proceedings of Eurographics/ IEEE-VGTC Symposium on Visualization, vol. 28, pp. 1007–1014 (2009)
19. Vucini, E.: On visualization and reconstruction from non-uniform point sets, Ph.D. thesis, Vienna University of Technology (2009)
20. Wagner, H., Chen, C., Vuçini, E.: Efficient computation of persistent homology for cubical data. In: Proceedings of TopoInVis 2011. Mathematics and Visualization. Springer (2012)
21. Weber, G.H., Bremer, P.-T., Day, M.S., Bell, J.B., Pascucci, V.: Feature tracking using reeb graphs. In: Proceedings of Topology-based Methods in Visualization (2009)

Parallel Skeletonizing of Digital Images by Using Cellular Automata

Francisco Peña-Cantillana[1], Ainhoa Berciano[2,3],
Daniel Díaz-Pernil[3], and Miguel A. Gutiérrez-Naranjo[1]

[1] Research Group on Natural Computing
Department of Computer Science and Artificial Intelligence
University of Seville, Spain
frapencan@gmail.com, magutier@us.es
[2] Departamento de Didáctica de la Matemática y de las Ciencias Experimentales
University of the Basque Country
ainhoa.berciano@ehu.es
[3] Research Group on Computational Topology and Applied Mathematics
Department of Applied Mathematics
University of Seville, Spain
sbdani@us.es

Abstract. Recent developments of computer architectures together with alternative formal descriptions provide new challenges in the study of digital Images. In this paper we present a new implementation of the Guo & Hall algorithm [8] for skeletonizing images based on Cellular Automata. The implementation is performed in a real-time parallel way by using the GPU architecture. We show also some experiments of skeletonizing traffic signals which illustrates its possible use in real life problems.

1 Introduction

A car at 120 km/h takes 6 seconds to travel 200 meters. Recognizing traffic signals in real time is a challenge in the automotive industry and represents an important step for automatic driving [13,14]. Moreover, changes in visibility due to the weather, lighting etc. make necessary to store an *scheme* of the signal, instead of a picture of it. The *skeleton* of the signal can be a good way to represent the signal in an abstract way. Skeletonizing such images efficiently is crucial in the process, since a car automatically driven must react in a very short space of time according to signals.

In this paper we propose a real-time parallel implementation of the Guo & Hall algorithm [8] for skeletonizing images by using a device architecture called CUDA™, (Compute Unified Device Architecture) [21]. CUDA™ is a general purpose parallel computing architecture that allows the parallel NVIDIA Graphics Processors Units (GPUs) to solve many complex computational problems in a more efficient way than on a CPU [15].

M. Ferri et al. (Eds.): CTIC 2012, LNCS 7309, pp. 39–48, 2012.

The algorithm has been adapted on the theoretical basis using one of the most known models of the Natural Computing, *Cellular Automata* (CA). Natural Computing studies computational paradigms inspired from physics, chemistry and biology [11]. It abstracts the way in which nature acts, providing ideas for new computing models. One of the main research lines in Natural Computing is Cellular Automata, introduced by John Von Neumann with the biological motivation of obtaining self-replicating artificial systems that are also computationally universal.

CA are decentralized discrete computational systems[1]. They consist of large numbers of simple identical components (cells) placed on an N-dimensional grid with local connectivity defining the *neighbourhood* of a cell. The cells are in one of a finite set of *states*. A discrete global clock is assumed and the cells change their states synchronously depending on their own state and the states of the neighbours, as determined by a local update rule. Technically, a CA consists of two components. The first one is a *cellular space*: a lattice of N identical finite-state machines (cells) each with an identical pattern of local connections to other cells, with boundary conditions if the lattice is finite. The second component is a set of transition rules that gives the update state of each cell.

These features make CA suitable for dealing with some problems in the analysis of digital images, where pixels are identified with cells and the changes in a cell depends on the current state plus the current state of its neighbours and all the changes can be made simultaneously [9,17]. Nonetheless, the inherent features of CA for dealing with Image Analysis have found the limits of sequential computers. The theoretical parallel framework of CA could not be efficiently implemented in one-processor computers. Recently, the development of new parallel architectures has brought a renewed interest in the use of CA for Image Analysis.

The paper is organised as follows: Firstly, we recall the basic notions of the Guo & Hall algorithm and give some ideas of our implementation on CUDA inspired on CA. In Section 3 we show illustrative examples of the use of our implementation and some comparisons with a sequential one. Finally, Section 4 is dedicated to conclusions and future work.

2   Guo and Hall Algorithm

Representing a shape with a small amount of information is a challenge in computer vision. Skeletonization is one of the approaches to this purpose, converting the initial image into a more compact representation and keeping the meaning features. The conversion should remove redundant information, but it should also keep the basic structure. Skeletonization is usually considered as a pre-process in pattern recognition algorithms, but its study is also interesting by itself for the analysis of line-based images as texts, line drawings, human fingerprints or cartography.

---

[1] We assume that the reader is familiar with the basic concepts of CA. More information in [10,20].

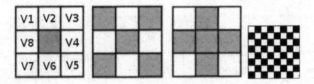

Fig. 1. (a) Order of neighbour pixels of P. (b) Sail configuration. (c) Cross configuration. (d) Chess configuration of an image.

The concept was introduced by Blum [5,6] under the name of medial axis transform. There are many algorithms published in this topic (see [16]) and there are many different approaches to the problem, among them the ones based on distance transform of the shape and skeleton pruning based on branch analysis[2].

In this paper, the skeleton is obtained by an iterative procedure of thinning: the border points are removed as long as they are not considered significant. The remaining set of points is called the *skeleton*. Among the parallel algorithms following this idea, special attention deserves the so-called 1-subcycle parallel algorithms or fully parallel algorithms [8]. We present a new implementation of the algorithm of Guo & Hall by using the new technology GPGPU. In this algorithm, the contour pixels are examined for deletion in an iterative process. The decision is based on a 3×3 neighbourhood. The image is divided into two disjoint areas (sub-sections), similarly to a chess board. The algorithm consists on two sub-iterations where the removal of redundant pixels from *white* and *black* sections are alternated. This is repeated until there are no redundant pixels left.

According to [7], given a pixel P, we will denote by $P_1, \ldots, P_8$ the clockwise enumeration of its eight neighbour pixels (see Fig. 1 a)) and P as a Boolean variable, with the truth value 1 if P is *black* and 0 if P is *white*. Two parameters are defined:

$$B(P) = \sum_{i=1}^{i=8} P_i$$
$$C(P) = (\neg P_2 \wedge (P_3 \vee P_4)) + (\neg P_4 \wedge (P_5 \vee P_6))$$
$$+ (\neg P_6 \wedge (P_7 \vee P_8)) + (\neg P_8 \wedge (P_1 \vee P_2))$$

where $B(P)$ is the number of black neighbour pixels, and $C(P)$ is the *connectivity operator* given by the number of white neighbour pixels of P where some of the next two pixels is black, following the order of pixels of Fig. 1 (a).

In each iteration, each pixel P is deleted (changed to white) if and only if all of the following conditions are satisfied:

1. $C(P) = 1$; this condition is necessary for preserving local connectivity when P is deleted.
2. $(P_1 \wedge P_3 \wedge P_5 \wedge P_7) \vee (P_2 \wedge P_4 \wedge P_6 \wedge P_8) = FALSE$; i.e., we cannot to find a configuration of neighbour pixels with the way of Fig. 1 b) and c)).
3. $B(P) > 1$.

---

[2] See, for example, [1,3,19,4,2].

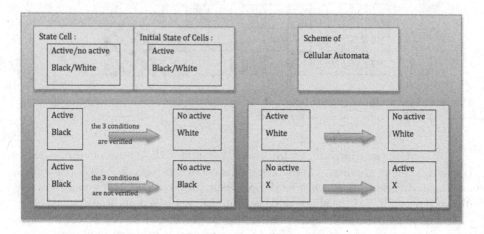

Fig. 2. Scheme of our Cellular Automata

2.1 CA Formal Framework

The formal description of the algorithm in the framework of CA requires to provide the cellular space and the set of rules.

- Given a $n \times m$ image, let us consider the CA *cellular space* on a rectangular grid $(n+2) \times (m+2)$ with 8-adjacency[3]. The set of possible states is $\{x, y\}$ for $x \in \{0, 1\}$, where 0 stands for white and 1 stands for black, and $y \in {0, 1}$ where 0 stands for active and 1 for no active. In the initial configuration, we will consider the central $n \times m$ grid, where the cells will take the initial state according with the color of the corresponding pixel in the image. The remaining one-width framework will have the initial state 0 (white). The initial value of y depends of the position of pixel in a image with the same size of the input image, but taking a chess configuration (see Fig. 1 (d)). So, if the pixel is black we consider the cell as active and if the pixel is white the cell is no active.
- For the description of the set of rules, it is necessary to provide the conditions necessary for the change in the state of a pixel. In this case, such conditions are taken directly from the Guo & Hall algorithm (see above). A cell will change its state if we can apply one of the four rules that appear in Fig. 2; for example, a cell changes its state from $\{1, 1\}$ to $\{0, 0\}$ (from {black,active} to {white,no active}) if it satisfies the conditions to be considered *redundant*, which includes its current state and the color of the pixels in its 8-neighbourhood.

---

[3] We consider this extra framework in order to avoid boundary conditions in the description of the algorithm.

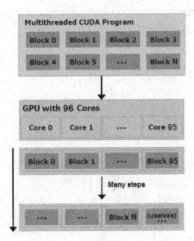

Fig. 3. Scheme of threads for our skeletonizing

2.2 Parallel Implementation

GPUs constitute nowadays a solid alternative for high performance computing, and the advent of CUDA™ allows programmers a friendly model to accelerate a broad range of applications. The parallel implementation of the Guo & Hall algorithm described above has been developed by using Microsoft Visual Studio 2008 Professional Edition (C++) with the plugging Parallel Nsight (CUDA™) under Microsoft Windows 7 Professional with 32 bits. CUDA™ C, an extension of C for implementations of executable kernels in parallel with graphical cards NVIDIA has been used to implement the CA. It has been necessary the *nvcc compiler* of CUDA™ Toolkit and some libraries from openCV to the treatment of input and output images.

The experiments have been performed on a computer with a CPU AMD Athlon II x4 645, which allows to work with four cores of 64 bits to 3.1 GHz. The computer has four blocks of 512KB of L2 cache memory and 4 GB DDR3 to 1600 MHz of main memory.

The used graphical card (GPU) is an NVIDIA Geforce GT240 composed by 12 *Stream Processors* with a total of 96 cores to 1340 MHz. It has 1 GB DDR3 main memory in a 128 bits bus to 700 MHz. So, the transfer rate obtained is by 54.4 Gbps. The used Constant Memory is 64 KB and the Shared Memory is 16 KB. Its Compute Capability level is 1.2 (from 1.0 to 2.1).

We can deal N blocks of threads for the complete image in our GPU of 96 cores, as we can see in Fig. 3. We need more threads than pixels if the height and width of the image are not multiples of 16, i.e., we can have useless threads. The Figure 4 shows a flowchart of the implementation on CUDA of the algorithm presented in this paper.

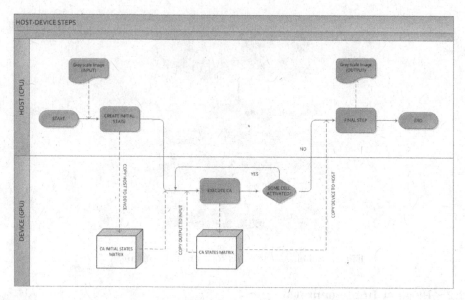

Fig. 4. Flowchart

3 Skeletonizing

In this section we show the results of some experiments of skeletonizing traffic signals with our parallel implementation of the Guo & Hall algorithm. Notice that, the algorithm is described for black and white colour images, so if the image given is a colour image or an image in grey scale, firstly we apply an algorithm of binarization, and later our parallel software.

Firstly, in Figure 5 we can see an example of skeletonizing some traffic signals. Notice that the skeletonizing of signals with the original foreground in black and the background white provides meaningful information about the message, but in the opposite way, the output of the skeletonizing is not a recognisable image and not suitable for automatic recognition.

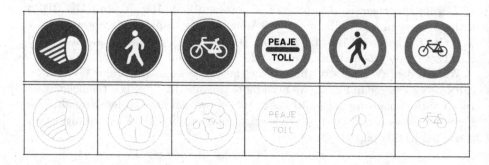

Fig. 5. Some traffic signals and their skeletonization

Fig. 6. A real photograph, its binarization, its skeletonizing, its inverse binarization and its inverse skeletonizing (from left to right)

Next, we provide an example of a realistic recognising problem. In Figure 6, we can see a photograph of size 789×1317. It has been binarized by using a threshold method by using a threshold 100 on a gray scale $0, \ldots, 255$. At the bottom of the Figure, we can see its skeletonizing and the skeletonizing of the inverse thresholding.

We finish this section by showing the results of some experiments performed with our implementation. We have taken 36 totally black images of $n \times n$ pixels[4], from $n = 125$ to $n = 4500$ with a regular increment of 125 pixels of side. Figure 7 (top) shows the time in milliseconds of the application of our implementation of the Guo & Hall algorithm in CA for 1, 30, 60 and 90 steps in the skeletonizing

---

[4] Theoretically, this is the worst case, since the time inverted by the algorithm depends on the size of the biggest black connected component of the original image.

process. Figure 7 (bottom) shows the same study for a sequential implementation of the algorithm. Finally, Figure 8 shows a comparison of our implementation vs. the sequential one by taking 90 steps in the Guo & Hall algorithm.

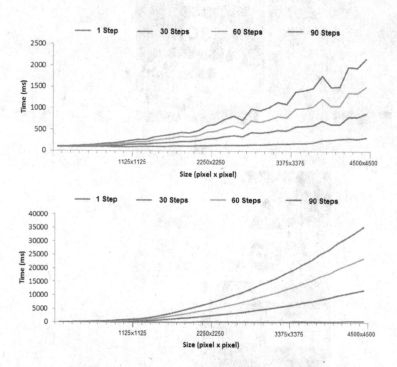

Fig. 7. Experimental time obtained for the Guo & Hall algorithm 36 totally black images of $n \times n$ pixels, from $n = 125$ to $n = 4500$ with a regular increment of 125 pixels of side. Top image shows the time of our parallel implementation in CA. Bottom image shows the time for a sequential implementation.

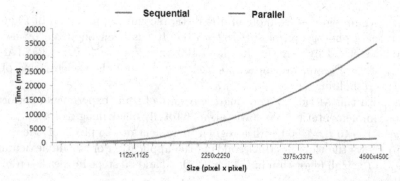

Fig. 8. A comparison of our implementation vs. the sequential one by taking 90 steps in the Guo & Hall algorithm

4 Conclusions

Computer vision is a hard task and a challenge in the next years. Classical sequential algorithms need to be revisited and adapted to the novel technologies, but the new developments also need the support of deep theoretical foundations. Using Cellular Automata is not a new concept in the study of digital images (see, for example, [18]) but the intrinsic parallelism of CA could not be effectively explored due to the limitations of sequential computers. Only recently, the new architectures of GPU has started to be studied as a tool for realistic implementations (see, e.g., [12]). This paper goes in this line by proposing a new implementation of one of the basic pre-processing problems for image analysis, the Guo & Hall algorithm. The quick development of new hardware architectures will provide in the next years a challenge for the effective parallel implementation of many other approaches in Image Analysis.

Acknowledgements. DDP and MAGN acknowledge the support of the projects TIN2008-04487-E and TIN-2009-13192 of the Ministerio de Ciencia e Innovación of Spain and the support of the Project of Excellence with *Investigador de Reconocida Valía* of the Junta de Andalucía, grant P08-TIC-04200. AB acknowledges the support of the project MTM2009-12716 of the Ministerio de Educación y Ciencia, the project EHU09/04 and the "Computational Topology and Applied Mathematics" PAICYT research group FQM-296.

References

1. Arcelli, C., di Baja, G.S.: Euclidean skeleton via centre-of-maximal-disc extraction. Image and Vision Computing 11(3), 163–173 (1993)
2. Attali, D., Boissonnat, J.D., Edelsbrunner, H.: Stability and Computation of Medial Axes - a State-of-the-Art Report. In: Mathematical Foundations of Scientific Visualization, Computer Graphics, and Massive Data Exploration, ch. 6, pp. 109–125. Springer, Heidelberg (2009)
3. di Baja, G.S., Thiel, E.: Skeletonization algorithm running on path-based distance maps. Image and Vision Computing 14(1), 47–57 (1996)
4. Biasotti, S., Attali, D., Boissonnat, J.-D., Edelsbrunner, H., Elber, G., Mortara, M., Baja, G.S., Spagnuolo, M., Tanase, M., Veltkamp, R.: Skeletal structures. In: Floriani, L., Spagnuolo, M. (eds.) Shape Analysis and Structuring. Mathematics and Visualization, pp. 145–183. Springer, Heidelberg (2008)
5. Blum, H.: An associative machine for dealing with the visual field and some of its biological implications. Computer and Mathematical Sciences Laboratory, Electronics Research Directorate, Air Force Cambridge Research Laboratories, Office of Aerospace Research, United States Air Force (1962)
6. Blum, H.: An associative machine for dealing with the visual field and some of its biological implications. In: Bernard, E.E., Kare, M.R. (eds.) Biological Prototypes and Synthetic Systems, vol. 1, pp. 244–260. Plenum Press, New York (1962); Proceedings of the 2nd Annual Bionics Symposium, held at Cornell University (1961)
7. Bräunl, T.: Parallel image processing. Springer (2001)
8. Guo, Z., Hall, R.W.: Parallel thinning with two-subiteration algorithms. Communications of the ACM 32, 359–373 (1989)

9. Hernandez, G., Herrmann, H.J.: Cellular-automata for elementary image-enhancement. Graphical Models and Image Processing 58(1), 82–89 (1996)
10. Kari, J.: Theory of cellular automata: A survey. Theoretical Computer Science 334(1-3), 3–33 (2005)
11. Kari, L., Rozenberg, G.: The many facets of natural computing. Communications of the ACM 51(10), 72–83 (2008)
12. Kauffmann, C., Piché, N.: A cellular automaton framework for image processing on GPU. In: Yin, P.Y. (ed.) Pattern Recoginition, pp. 353–375. InTech (2009)
13. Klette, R., Ahn, J., Haeusler, R., Herman, S., Huang, J., Khan, W., Manoharan, S., Morales, S., Morris, J., Nicolescu, R., Ren, F., Schauwecker, K., Yang, X.: Advance in vision-based driver assistance. In: 2011 International Conference on Electric Technology and Civil Engineering (ICETCE), pp. 987–990 (April 2011)
14. Mohapatra, A.G.: Computer vision based smart lane departure warning system for vehicle dynamics control. Sensors & Transducers Journal 132(9), 122–135 (2011)
15. Owens, J.D., Luebke, D., Govindaraju, N., Harris, M., Krüger, J., Lefohn, A., Purcell, T.J.: A survey of general-purpose computation on graphics hardware. Computer Graphics Forum 26(1), 80–113 (2007)
16. Saeed, K., Tabedzki, M., Rybnik, M., Adamski, M.: K3M: A universal algorithm for image skeletonization and a review of thinning techniques. Applied Mathematics and Computer Science 20(2), 317–335 (2010)
17. de Saint Pierre, T., Milgram, M.: New and efficient cellular algorithms for image processing. CVGIP: Image Understanding 55(3), 261–274 (1992)
18. Selvapeter, P.J., Hordijk, W.: Cellular automata for image noise filtering. In: NaBIC, pp. 193–197. IEEE (2009)
19. Siddiqi, K., Pizer, S.M.: Medial representations: mathematics, algorithms and applications. In: Computational Imaging and Vision. Springer (2008)
20. Wolfram, S.: Cellular Automata and Complexity: Collected Papers. Perseus Books Group (1994)
21. NVIDIA Corporation. NVIDIA CUDA^{tm} Programming Guide,
    http://www.nvidia.com/object/cuda_home_new.html

Towards a Certified Computation of Homology Groups for Digital Images*

Jónathan Heras[1], Maxime Dénès[2], Gadea Mata[1], Anders Mörtberg[3],
María Poza[1], and Vincent Siles[3]

[1] Department of Mathematics and Computer Science of University of La Rioja
[2] INRIA Sophia Antipolis, Méditerranée
[3] University of Gothenburg
{jonathan.heras,gadea.mata,maria.poza}@unirioja.es,
Maxime.Denes@inria.fr, {mortberg,siles}@chalmers.se

Abstract. In this paper we report on a project to obtain a verified computation of homology groups of digital images. The methodology is based on programming and executing *inside* the CoQ proof assistant. Though more research is needed to integrate and make efficient more processing tools, we present some examples partially computed in CoQ from real biomedical images.

Keywords: Homology, Discrete Morse Theory, Proof assistant tools, Coq, SSReflect, Synapses.

1 Introduction

The discipline of Algebraic Digital Topology, or more specifically, the computation of homology groups from digital images is mature enough (see, for instance, [27], one among many good references) to go one step further and investigate the possibility of a *certified computation* (i.e., formally verified by proving correctness using an *interactive* proof assistant) in digital topology, as it happens in other areas of computer mathematics (see [8]).

In a very rough manner, the process to be verified is reflected in Figure 1. Putting it into words, from the black pixels of a monochromatic image a simplicial complex is obtained (by means of a triangulation procedure); subsequently, from the simplicial complex, its *boundary (or incidence) matrices* are constructed, and finally, *homology* can be computed. If we work with coefficients over a field (and it is well-known that it is enough to take as coefficients the field $\mathbb{Z}/2\mathbb{Z}$, when we work with 2D and 3D digital images) and if only the *dimensions* of the homology groups (as vector spaces) are looked for, then having a program able to compute the rank of a matrix is sufficient to accomplish the whole task.

---

* Partially supported by Ministerio de Educación y Ciencia, project MTM2009-13842-C02-01, and by the European Union's 7th Framework Programme under grant agreement nr. 243847 (ForMath).

M. Ferri et al. (Eds.): CTIC 2012, LNCS 7309, pp. 49–57, 2012.

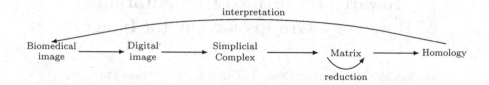

Fig. 1. Computing homology from a digital image

This architecture is particularized in this paper with a real problem that appeared in an industrial application and with the CoQ proof assistant as programming and verifying tool.

The rest of this paper is organized as follows. Section 2 is devoted to present an example, coming from the biomedical context, as a test-case for our formal development. The formalization process is explained in Section 3, focusing on the link between boundary matrices and homology groups. Section 4 explains how the certified programs can be used to effectively compute homology of images. A way to deal with the management of the huge matrices produced by biomedical images is presented in Section 5. The paper ends with a section of Conclusions and Further work, and the bibliography.

2   Motivation

When developing formal proofs, a major issue is ensuring that concepts are defined in a way that will be applicable to concrete use. In our case, we are developing a general theory of effective simplicial homology as part of the Formath project [1]. We decided to validate our design choices on biomedical digital images obtained from synaptical structures.

Synapses are the points of connection between neurons. The relevance of synapses comes from the fact that they are related to the computational capabilities of the brain.

The possibility of changing the number of synapses may be an important asset in the treatment of neurological diseases, such as Alzheimer, see [26]. Therefore, we can claim that an efficient, reliable and automatic method for counting synapses is instrumental in the study of the evolution of synapses in scientific experiments.

Up to now, the method to count synapses was manual, see [6]. This was impractical since it implies a considerable time investment. In order to improve this process, a plug-in called SynapCountJ [17] for the ImageJ environment [22] has been developed.

The procedure implemented in this software to handle neuron images can be split into two steps. First, taking as input three images of a neuron, namely the neuron with two different antibody markers and the structure of the neuron, SynapCountJ produces a bitmap where synapses are the connected components, see Figure 2 (the same images with higher resolution are accessible at http://www.unirioja.es/cu/joheras/synapses/). Then the second step consists in counting the connected components of the bitmap. A detailed explanation of the procedure was given in [13].

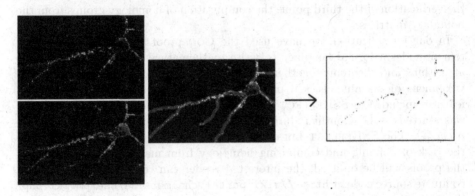

Fig. 2. Example of the results produced by SynapCountJ

To test the suitability of this program, biologists consider, on the one hand, control cultures and, on the other hand, cultures under the effect of some drugs; in this way, the evolution of the density of the occurrence of synapses under the effect of those drugs can be determined. For instance, using the chemical inhibitor GSK3, the evolution percentage manually obtained is 36% and the one obtained with SynapCountJ is 36.6%. Thus, the experimental results obtained with SynapCountJ were considered (by the biologists) very satisfactory.

The former step of the procedure implemented in SynapCountJ, the extraction of a bitmap with the synapses from three images of the neurons, is carried out based on solid previous experience of experimental scientists; therefore, they consider it as a safe process. The latter step, the computation of connected components, can be solved with many algorithms and is an interesting test case for our framework where we can compute the homology in dimension 0 of such images. This is a well known procedure to measure the amount of connected components of an image, even if more elementary methods are also applicable.

3   Verification in COQ/SSREFLECT

In the introduction we have explained a method, based on simplicial homology, to study the homology of a digital image which consists of: (1) building a simplicial complex from the image, (2) generating the boundary matrices associated with the simplicial complex, and (3) computing the homology from the boundary matrices. Notwithstanding that cubical complexes are more suitable to encode monochromatic images, it is worth noting that we are working on top of previous formalization efforts which deal with simplicial complexes.

The correctness of the programs in charge of both the construction of a simplicial complex from an image and the generation of the boundary matrices associated with a simplicial complex have been formally proved using proof assistant tools as can be seen in [21] and [14] respectively. Then, there only remains the verification of the third point, the computation of homology groups from the boundary matrices.

In our formalization, we have used the Coq proof assistant [5]. This system provides a formal language to write mathematical definitions, executable algorithms and theorems together with an environment for semi-interactive development of machine-checked proofs. In addition, we take advantage of the features included in SSReflect [9], an extension for Coq whose development was started by G. Gonthier during the formal proof of the *Four Color Theorem* [8]. The SSReflect libraries include enough ingredients to undertake the task of defining and computing homology from matrices. Some details of the proofs will be omitted; the interested reader can consult the original and complete source code at http://wiki.portal.chalmers.se/cse/pmwiki.php/ForMath/ProofExamples.

First of all, we define the notion of homology in Coq. Let K be a field, $V1, V2, V3$ vector spaces on K, and $f : V1 \rightarrow V2, g : V2 \rightarrow V3$ linear applications; then, the *Homology of f, g* is the quotient between the *kernel* of g and the *image* of f. This is translated into Coq in the following way.

```
Variable (K : fieldType) (V1 V2 V3 : vectType K)
         (f : linearApp V1 V2) (g : linearApp V2 V3).
Definition Homology := ((lker g) :\: (limg f)).
```

Nevertheless, we do not usually work with linear applications when trying to compute homology but with the matrices representing those linear applications. In particular, as we are working on a field K, given two matrices with coefficients in this field, let us called them, mxf and mxg of sizes $v1 \times v2$ and $v2 \times v3$ respectively and such that their product is the null matrix, the dimension of the corresponding homology vector space is given by the formula: $v2 - rank(mxg) - rank(mxf)$. This definition is introduced in Coq as follows.

```
Definition dim_homology (mxf:'M[K]_(v1,v2)) (mxg:'M[K]_(v2,v3)) :=
    v2 - \rank mxg - \rank mxf.
```

Now, the correctness of dim_homology can be shown by proving that given two matrices mxf and mxg whose product is the null matrix (mxf *m mxg = 0), then the result obtained using dim_homology is the dimension of the homology group associated with the linear applications defined from mxf and mxg ((LinearApp mxf) and (LinearApp mxg)).

```
Lemma dimHomologyrankE: mxf *m mxg = 0 ->
    \dim Homology (LinearApp mxf) (LinearApp mxg) =
    dim_homology mxf mxg.
```

However the use of SSReflect libraries may trigger heavy computations during deduction steps, that would not terminate within a reasonable amount of time.

To handle this issue, some definitions like matrices are locked in a way that do not allow direct computations.

To overcome this pitfall, we use the matrix representation and the rank algorithm developed in [4] to define ex_homology which takes as argument two such matrices (represented by means of lists of lists) mxf and mxg which dimensions are v1×v2 and v2×v3 respectively, and computes the homology.

```
Definition ex_homology (v1 v2 v3:nat) (mxf mxg : seqMatrix K) :=
    v2 - (rank v2 v3 mxg) - (rank v1 v2 mxf).
```

Finally, we prove the correctness of ex_homology by showing its equivalence to dim_homology up to a change of representation (this domain transformation is given by seqmx_of_mx).

```
Lemma ex_homology_rankE: forall (mxf: 'M[K]_(w1,w2)) (mxg : 'M[K]_
    (w2,w3)), ex_homology (seqmx_of_mx mxf) (seqmx_of_mx mxg) =
    dim_homology mxf mxg.
```

Then, we have an executable program to compute homology, for any dimension, whose correctness has been verified in CoQ; therefore, we can claim that its results will always be correct.

4   Computing Homology with CoQ

An example is presented in this section in order to clarify how we can compute homology groups in CoQ. Let us consider the simplicial complex of the left side of Figure 3. If we impose a lexicographical order on the simplices of the same dimension of this simplicial complex, its boundary matrix in dimension 1 is the one presented in the right side of Figure 3; it is worth noting that the rest of boundary matrices are empty, in particular we do not consider the empty set as an element of dimension −1.

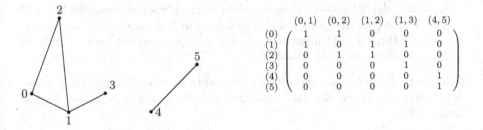

Fig. 3. Simplicial complex and its boundary matrix

The procedure to compute the homology (note that it only makes sense to compute homology in dimensions 0 and 1) of the simplicial complex of Figure 3 is as follows. Firstly, we define the boundary matrices.

```
Definition d0_ex1 := [::].
Definition d1_ex1 := [::[::1;1;0;0;0];
                        [::1;0;1;1;0];
                        [::0;1;1;0;0];
                        [::0;0;0;1;0];
                        [::0;0;0;0;1];
                        [::0;0;0;0;1]].
Definition d2_ex1 := [::].
```

Eventually, we can compute the homology using the following instructions.

```
Eval vm_compute in (ex_homology 0 6 5 d0_ex1 d1_ex1).
Eval vm_compute in (ex_homology 6 5 0 d1_ex1 d2_ex1).
```

obtaining 2 and 1 respectively. In the same way, we could compute homology from the boundary matrices associated with the simplicial complex generated from a digital image. However, if we try to compute the homology from the images produced by SynapCountJ (see Figure 2), COQ is not able to handle those images yet, due to the size of data involved.

It is worth noting that COQ is a Proof Assistant and not a Computer Algebra system. Efficient implementations of mathematical algorithms running inside COQ is an ongoing effort, as shown by recent works on efficient real numbers [16], machine integers and arrays [2] or a previous approach to compiled execution of internal computations [10].

We devise a couple of ways to achieve better efficiency:

- Improve the runtime system using the extraction mechanism which translates COQ code to a functional programming language like OCAML or Haskell. However, this would not allow us to reuse the result of our homological computations for further proofs. Indeed, output of external programs are untrusted so they cannot be imported. Instead, we are using a recent intermediate approach consisting in internally compiling COQ terms to OCAML with performance comparable to extracted code [18].
- Optimize algorithms and representations using sparse matrices, which is well suited to simplicial complexes obtained from digital images. We have developed an Haskell implementation of such an algorithm but we still need to formally verify its correctness.

In the next section we describe another method to overcome the efficiency drawback, based on reducing the size of matrices while keeping the same homological information.

5   Computing Discrete Vector Fields

The method that we are using for the reduction process is based on *Discrete Morse Theory* [7]; namely, we work in the algebraic setting of this theory which was described in [25]. Roughly speaking, the aim of Discrete Morse Theory consists of finding *simplicial collapses* which transform a simplicial complex $\mathcal{K}$ into a smaller one but keeping its homological properties. In this context, the instrumental tool are *admissible discrete vector fields* which allows one to reduce the amount of information removing "useless" information but keeping the homological properties of the original object.

The use of these techniques from Discrete Morse Theory has been welcomed in the study of homological properties of digital images, see [3,11,15], for instance. This is due to the fact that the size of the cellular object associated with an image can be huge, but the choice of an appropriate vector field can produce a much smaller object.

So, the question now is given a cellular complex how we can produce a vector field as large as possible (the larger the vector field, the smaller the reduced object). Several approaches to solve this problem have been studied as can be seen in [24,12,23,19], the strategy that we have chosen was explained in [25]. It is not the aim of this paper to describe that algorithm (from now on, called RS's algorithm; RS stands for Romero–Sergeraert); but, we just introduce some ideas. This algorithm takes as input one of the boundary matrices associated with the cellular complex and provides an admissible discrete vector field (subsequently, from the matrix and the vector field a reduced matrix can be obtained).

The algorithm has been implemented in Haskell; and, some remarkable results have been obtained in the reduction process. As benchmark to test our programs, we have considered matrices coming from, on the one hand, 500 randomly generated images; and, on the other hand, biomedical images. In the former case, the size of the matrices was initially around 100×300, and after the reduction process the average size was 5×50. Using the original matrices CoQ takes around 12 seconds to compute their rank; on the contrary, using the reduced matrices CoQ only needs milliseconds. In the latter case, the matrices coming from biomedical images, the size of matrices is reduced from around 690×1400 to 97×500. In this case, CoQ cannot deal with the original matrices; on the contrary, it is able to handle matrices as the ones obtained after applying the reduction programs and compute the results in, approximately, 25 seconds.

As a final remark, let us explain the main reason for using Haskell to implement the RS algorithm. The use of this language is due to the fact that Haskell is quite close to CoQ; and, therefore, algorithms implemented in Haskell can be verified using CoQ, a question which is, as we have seen, instrumental in our developments. In particular, the formalization of the correctness of the algorithm in charge of constructing an admissible discrete vector field given a matrix is ongoing work; and, up to now, we have certified that our programs build a discrete vector field. The proof of the admissibility property remains as further work.

6   Conclusions and Further Work

In this paper, we have presented how we can use Algebraic Topology techniques to study biomedical images in a reliable manner. The first step consists in processing the biomedical images to obtain an image where homological information is as explicit as possible. Subsequently, using programs whose correctness has been verified in the COQ/SSREFLECT proof assistant, homological properties from the pre-processed image are obtained, which in turn are interpreted as features of the original image.

This methodology has been applied in this paper to the problem of determining the number of synapses of a neuron. In this case, the problem is reduced to measure the number of connected components of a monochromatic image. An issue which can be solved, even if it is not the straightforward manner, thanks to the computation of the homology group in dimension 0 of the image.

The use of certified tools able to compute homology groups will be important in the future; for instance, to recognize the structure of a neuron; a problem which seems to involve the homology group in dimension 1, see [20]. Other techniques, like the ones of persistent homology, could be applied in stacks of neurons to remove the noise of the images and help to the detection of the dendrites (the branches of the neuron).

Some formalization aspects also remain as future work. We have already mentioned the on-going work around proving the correctness of the admissible discrete vector fields programs. Moreover, certifying the correctness of integer homology computation is also further work (some results about the formalization of the Smith Normal Form are already encoded in COQ, see [4]).

As we previously mentioned, we are still working on efficiency issues but switching to better representations and more efficient algorithms will not require to redo the proofs related to homology.

References

1. ForMath: formalisation of mathematics,
   http://wiki.portal.chalmers.se/cse/pmwiki.php/ForMath/ForMath
2. Armand, M., Grégoire, B., Spiwack, A., Théry, L.: Extending COQ with Imperative Features and Its Application to SAT Verification. In: Kaufmann, M., Paulson, L.C. (eds.) ITP 2010. LNCS, vol. 6172, pp. 83–98. Springer, Heidelberg (2010)
3. Cazals, F., Chazal, F., Lewiner, T.: Molecular shape analysis based upon Morse-Smale complex and the Connolly function. In: Proceedings 19th ACM Symposium on Computational Geometry (SCG 2003), pp. 351–360 (2003)
4. Cohen, C., Dénès, M., Mörtberg, A., Siles, V.: Smith Normal form and executable rank for matrices,
   http://wiki.portal.chalmers.se/cse/pmwiki.php/ForMath/ProofExamples
5. COQ development team. The Coq Proof Assistant Reference Manual, version 8.3. Technical report (2010)
6. Cuesto, G., et al.: Phosphoinositide-3-Kinase Activation Controls Synaptogenesis and Spinogenesis in Hippocampal Neurons. The Journal of Neuroscience 31(8), 2721–2733 (2011)
7. Forman, R.: Morse theory for cell complexes. Advances in Mathematics 134, 90–145 (1998)
8. Gonthier, G.: Formal proof - The Four-Color Theorem, vol. 55. Notices of the American Mathematical Society (2008)
9. Gonthier, G., Mahboubi, A.: A Small Scale Reflection Extension for the Coq system. Technical report, Microsoft Research INRIA (2009),
   http://hal.inria.fr/inria-00258384

10. Grégoire, B., Leroy, X.: A compiled implementation of strong reduction. In: Proceedings of the Seventh ACM SIGPLAN international Conference on Functional Programming, ICFP 2002, pp. 235–246. ACM, New York (2002)
11. Gyulassy, A., Bremer, P., Hamann, B., Pascucci, V.: A practical approach to Morse-Smale complex computation: Scalability and generality. IEEE Transactions on Visualization and Computer Graphics 14(6), 1619–1626 (2008)
12. Harker, S., et al.: The Efficiency of a Homology Algorithm based on Discrete Morse Theory and Coreductions. In: Proceedings 3rd International Workshop on Computational Topology in Image Context (CTIC 2010). Image A, vol. 1, pp. 41–47 (2010)
13. Heras, J., Mata, G., Poza, M., Rubio, J.: Homological processing of biomedical digital images: automation and certification. Technical report (2010), http://wiki.portal.chalmers.se/cse/uploads/ForMath/hpbdiac
14. Heras, J., Poza, M., Dénès, M., Rideau, L.: Incidence Simplicial Matrices Formalized in Coq/SSReflect. In: Davenport, J.H., Farmer, W.M., Urban, J., Rabe, F. (eds.) Calculemus/ MKM 2011. LNCS (LNAI), vol. 6824, pp. 30–44. Springer, Heidelberg (2011)
15. Jerse, G., Kosta, N.M.: Tracking features in image sequences using discrete Morse functions. In: Proceedings 3rd International Workshop on Computational Topology in Image Context (CTIC 2010). Image A, vol. 1, pp. 27–32 (2010)
16. Krebbers, R., Spitters, B.: Computer Certified Efficient Exact Reals in Coq. In: Davenport, J.H., Farmer, W.M., Urban, J., Rabe, F. (eds.) Calculemus/MKM 2011. LNCS (LNAI), vol. 6824, pp. 90–106. Springer, Heidelberg (2011)
17. Mata, G.: SynapsCountJ. University of La Rioja (2011), http://imagejdocu.tudor.lu/doku.php?id=plugin:utilities:synapsescountj:start
18. Boespflug, M., Dénès, M., Grégoire, B.: Full Reduction at Full Throttle. In: Jouannaud, J.-P., Shao, Z. (eds.) CPP 2011. LNCS, vol. 7086, pp. 362–377. Springer, Heidelberg (2011)
19. Molina-Abril, H., Real, P.: A Homological–Based Description of Subdivided nD Objects. In: Real, P., Diaz-Pernil, D., Molina-Abril, H., Berciano, A., Kropatsch, W. (eds.) CAIP 2011, Part I. LNCS, vol. 6854, pp. 42–50. Springer, Heidelberg (2011)
20. Mrozek, M., et al.: Homological methods for extraction and analysis of linear features in multidimensional images. Pattern Recognition 45(1), 285–298 (2012)
21. F.L.R.: node. From Digital Images to Simplicial Complexes: A report. Technical report (2011), http://wiki.portal.chalmers.se/cse/uploads/ForMath/fditscr
22. Rasband, W.S.: ImageJ: Image Processing and Analysis in Java (2003), http://rsb.info.nih.gov/ij/
23. Real, P., Molina-Abril, H.: Towards Optimality in Discrete Morse Theory through Chain Homotopies. In: Proceedings 3rd International Workshop on Computational Topology in Image Context (CTIC 2010). Image A, vol. 1, pp. 33–40 (2010)
24. Robins, V., Wood, P., Sheppard, A.: Theory and algorithms for constructing discrete Morse complexes from grayscale digital images. IEEE Transactions on Pattern Analysis and Machine Intelligence 33(8), 1646–1658 (2011)
25. Romero, A., Sergeraert, F.: Discrete Vector Fields and Fundamental Algebraic Topology (2010), http://arxiv.org/abs/1005.5685v1
26. Selkoe, D.J.: Alzheimer's disease is a synaptic failure. Science 298(5594), 789–791 (2002)
27. Ziou, D., Allili, M.: Generating Cubical Complexes from Image Data and Computation of the Euler number. Pattern Recognition 35, 2833–2839 (2002)

An Efficient Algorithm
to Compute Subsets of Points in $\mathbb{Z}^n$

Ana Pacheco and Pedro Real

Dpto. Matemática Aplicada I, ETS Ingeniería Informática, Universidad de Sevilla
{ampm,real}@us.es

Abstract. In this paper we show a more efficient algorithm than that in [8] to compute subsets of points non-congruent by isometries. This algorithm can be used to reconstruct the object from the digital image. Both algorithms are compared, highlighting the improvements obtained in terms of CPU time.

Keywords: digital image, grid, hypercube, isometry, n–xel.

1 Introduction

An n–*dimensional digital image* is a data structure typically representing a grid made up by a finite set of n–dimensional color hypercubes. The n–dimensional hypercubes of the grid are called n–xels for digital images of dimension n; particularly, pixels for $n = 2$ and voxels for $n = 3$.

By considering the central point of each n–dimensional hypercube of the grid, we construct a dual grid made up by n–dimensional hypercubes whose vertices are the central points of the hypercubes of the original grid.

In this way, the n–xels of an image are identified with vertices of n–dimensional hypercubes of the dual grid.

In Figure 1 we show more details about this construction for 2–dimensional binary digital images.

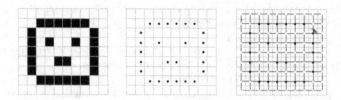

Fig. 1. From left to right: a binary digital image in a grid of size 10×9; central points of the image pixels; dual grid whose vertices are the central points of the squares of the original grid

In this sense, to represent images by using computational techniques it is necessary to fix a grid and the relations between the points.

M. Ferri et al. (Eds.): CTIC 2012, LNCS 7309, pp. 58–67, 2012.

Binary images are derived from a subdivision of the n–dimensional space into unit hypercubes of dimension n which intersect two by two in a hypercube of dimension $n - 1$. This subdivided space is equivalent to use as grid the n–dimensional discrete space $\mathbb{Z}^n$. The elements $(i_1, ..., i_n) \in \mathbb{Z}^n$ are the lattice points. Once the grid has been established, it is necessary to fix the neighborhood relations between the lattice points.

For a given lattice point, a *neighborhood* is defined typically by using a distance metric (see [3]). More concretely, two lattice points in $\mathbb{Z}^n$ are *neighboring points* if they are less than *epsilon* distance away. Depending on the values of epsilon, different types of neighborhoods can be defined.

For instance, Kong and Roscoe [7] defined three standard types of neighborhood in the three-dimensional space $\mathbb{Z}^3$: the *6–neighborhood*, the *18–neighborhood* and the *26–neighborhood*. These definitions are essentially equivalent to the corresponding definitions in Rosenfeld [12]. In Figure 2 these three types of neighborhood in $\mathbb{Z}^3$ are shown.

Fig. 2. A point $P \in \mathbb{Z}^3$ with each one of its: (a) six neighboring points satisfies $d_1(P,Q) = 1$; (b) eighteen neighboring points satisfies $d_1(P,Q) = 1$ or $d_1(P,Q) = 2$; and (c) twenty-six neighboring points satisfies $d_1(P,Q) = 1$, $d_1(P,Q) = 2$ or $d_1(P,Q) = 3$

For instance, a point $P \in \mathbb{Z}^4$ with each one of its: (a) eight neighboring points satisfies $d_1(P,Q) = 1$; (b) thirty-two neighboring points satisfies $d_1(P,Q) = 1$ or $d_1(P,Q) = 2$; (c) sixty-four neighboring points satisfies $d_1(P,Q) = 1$, $d_1(P,Q) = 2$ or $d_1(P,Q) = 3$; and (d) eighty neighboring points satisfies $d_1(P,Q) = 1$, $d_1(P,Q) = 2$, $d_1(P,Q) = 3$ or $d_1(P,Q) = 4$. See [5] for more details.

2 Preliminaries

In this section, we recall some basic notions about algebraic-topology, geometry, graph theory and digital images in order to do more understandable the paper.

Given a set S, an *order relation* on S is a relation $\preceq$ such that, for every $a, b, c \in S$ is held: (1) either $a \preceq b$, or $b \preceq a$; (2) if $a \preceq b$ and $b \preceq c$, then $a \preceq c$; (3) if $a \preceq b$ and $b \preceq a$, then $a = b$. Moreover, S is called *ordered set*. The *reverse order relation* $\succeq$ is the relation given by $a \succeq b$ if $b \preceq a$. Given two ordered sets S_1 and S_2, the *lexicographic order* on the Cartesian product $S_1 \times S_2$ is defined as $(a,b) \preceq (a',b')$ if and only if $a \prec a'$, or $a = a'$ and $b \preceq b'$.

A *distance* is a function $d : \mathbb{R}^n \times \mathbb{R}^n \rightarrow \mathbb{R}$ satisfying the following properties: (a) $d(x, y) \geq 0$; (b) $d(x, y) = 0$ if and only if $x = y$; (c) $d(x, y) = d(y, x)$ and (d) $d(x, z) \leq d(x, y) + d(y, z)$.

Some well-known distances are: (1) $d_1(x, y) = \sum_{i=1}^{n} |x_i - y_i|$; (2) $d_2(x, y) = \sqrt{\sum_{i=1}^{n}(x_i - y_i)^2}$; and $d_\infty(x, y) = max_{i=1}^{n}\{|x_i - y_i|\}$.

An *n–polytope* is the closure of an *n*–cell with flat faces. Particularly, a *polygon* is a 2–polytope and a *polyhedron* is a 3–polytope.

An *n–dimensional hypercube* (or *hypercube of dimension n*) is an *n*–polytope of 2^n vertices which satisfy certain distance conditions. Particularly, 2–dimensional and 3–dimensional hypercubes are called *squares* and *cubes*, respectively.

A map $f : X \rightarrow Y$ is called *isometry* if for any $a, b \in X$ is satisfied $d(f(a), f(b)) = d(a, b)$. Two objects O, O' are called *isometric* (or *congruent by isometries*) if there exists a bijective isometry from O to O'.

The *group of isometries of a cube* are the rigid motions which leave the cube invariant. This group has 48 elements.

The *group of isometries of a 4–dimensional hypercube* are the rigid motions which leave the hypercube invariant. This group has 384 elements (see [10]).

A *graph* $G = (V(G), E(G))$ consists of two finite sets: $V(G)$, the *vertex set* of the graph, which is a nonempty set of elements called *vertices*. $E(G)$, the *edge set* of the graph, which is a possibly empty set of elements called *edges*, such that every edge $e \in E(G)$ is assigned an unordered pair of vertices $\{u, v\}$, $(u \neq v)$ called the *end-vertices* of e, and e is said to join u and v. If there exists more than one edge between each pair of vertices, the graph is called *multi-graph*.

A *subgraph* of a graph G is a graph having all its vertices and edges in G.

Two graphs G and G' are called *isomorphic graphs* if there exists an isomorphism (bijective morphism) between them.

Let G be a multi-graph of vertices $v_1, v_2, \ldots, v_n$. The *adjacency matrix* of G is a $n \times n$ matrix $M(G) = (m_{ij})$ where the element m_{ij} is given by the number of edges which join the vertex v_i to the vertex v_j.

A *n–dimensional digital image* is a representation of an image of dimension *n* as a finite set of digital values, called *picture elements* or *n–xels*. Particularly, these elements are called *pixels* and *voxels* for digital images of dimension 2 and 3, respectively. Moreover, if the set of digital values is $\{0, 1\}$ then the image is called *binary digital image*.

3 Computing Subsets of Points in $\mathbb{Z}^n$

The first stage of this section consists in constructing subsets of points starting from the vertices of an *n*–dimensional unit hypercube. Then, the congruent ones by isometries of the *n*–dimensional space are ignored.

We assume that all the points in $\mathbb{Z}^n$ are assigned binary values, one or zero. The points whose value is 1 (resp. 0) are called 1–points (resp. 0–points). Given a finite subset of points, V, constructed starting from the vertices of an *n*–dimensional unit hypercube, we also assume that the points in V have a value of 1 while the points in the complement of V have a value of 0.

These subsets of points are determined as follows: the n–dimensional unit hypercube has 2^n vertices and each one of them can be a 1–point or a 0–point, so there exist 2^{2^n} subsets of points which can be constructed starting from the vertices of the n–dimensional unit hypercube. More concretely, there exist $C(2^n, c)$ subsets with $0 \leq c \leq 2^n$ 1–points. By using properties of combinatorial numbers, $C(2^n, 2^n - c) = C(2^n, c)$ is also the number of subsets with $2^n - c$ 1–points. In this way, the number of subsets with c 1–points is the same as the number of subsets with $2^n - c$ 1–points..

Below, we show the extension of the method shown in [8] to ignore congruent subsets that differ by isometries of the n–dimensional space. This method consisted in associating each subset with a multi-graph. The vertices of the multi-graph were the points of the subset and the number of edges between each pair of vertices u, v was determined by the square of Euclidean distance between u, v.

By considering the previous association between multi-graphs and subsets of points of the n–dimensional unit hypercube, it was natural to identify subsets with their respective associated multi-graphs.

A similar proof of Theorem 1 in [8] shows that two isomorphic subsets with at least 2^{n-1} points are isometric. The converse implication is obvious, taking into account that isometry is a stronger concept than isomorphism.

By considering previous results, Algorithm 1 in [8] (whose pseudocode is extended in Algorithm 3.1 for any dimension) was implemented.

Algorithm 3.1

Input: set of vertices of the n–dimensional unit hypercube with an order relation $\prec$.
// V: empty list to save the vertices of the non-isomorphic multi-graphs.
Output: non-congruent subsets by isometries of the n–dimensional space.
begin
    for $c = 2^{n-1}, ..., 2^n$ **do**
        Construct an ordered set $(V_c, \prec)$ containing to the $C(2^n, c)$ subsets with c 1–points
        for $(V_c)_i \subset V_c$ **do**
            Determine the multi-graph associated with $(V_c)_i$, $(G_c)_i$, whose adjacency matrix is $M_{(G_c)_i} = ((m_{(G_c)_i})_{pq})$ where $(m_{(G_c)_i})_{pq} = a_{pq}$, being a_{pq} the square Euclidean distance between v_p, v_q
            while $(V_c)_{i_1} \subset V_c$ & $(V_c)_{i_2} \subset V_c$ & $(V_c)_{i_1} \prec (V_c)_{i_2}$ **do**
                if $(G_c)_{i_1}$ and $(G_c)_{i_2}$ are isomorphic **then**
                    $(V_c)_{i_1}$ and $(V_c)_{i_2}$ are congruent by isometries
                    $V_c = V_c - \{(V_c)_{i_2}\}$
                end if
            end while
        end for
        $V = V \bigcup V_c$
    end for
    return V
end

By using as input the set of vertices of the n–dimensional unit hypercube arranged on an order relation $\prec$, for $2^{n-1} \leq c \leq 2^n$, this algorithm: (a) constructs an ordered set, $(V_c, \prec)$, containing to the $C(2^n, c)$ subsets with c 1–points; (b) associates each subset $(V_c)_i \subset V_c$ with a multi-graph $(G_c)_i$. The vertices of $(G_c)_i$ are the points of $(V_c)_i$ and the element a_{pq} of the adjacency matrix of $(G_c)_i$ corresponds with the square of Euclidean distance between v_p, v_q; and (c) checks if there exists an isomorphism between each pair of multi-graphs $(G_c)_{i_1}, (G_c)_{i_2}$, associated with two subsets $(V_c)_{i_1}, (V_c)_{i_2} \subset V_c$ which satisfy $(V_c)_{i_1} \prec (V_c)_{i_2}$.

Algorithm 3.1 allows us to ignore the subsets with $2^{n-1} \leq c \leq 2^n$ points obtained by isometries of the n–dimensional space. The subsets with $0 \leq c < 2^{n-1}$ points are determined by complementation.

Remark 1. Algorithm 3.1 only constructs subsets of vertices of the n–dimensional unit hypercube with at least 2^{n-1} points.

Remark 2. Given an order relation, $\prec$, on the vertices of the n–dimensional unit hypercube, Algorithm 3.1 determines the smallest non-congruent subsets with respect to $\prec$. Moreover, by changing the order relation, subsets congruent with these ones are obtained.

By using Algorithm 3.1, the following results can be proved.

Theorem 1. *In $\mathbb{Z}^3$, there exist (up to isometry): (a) six subsets with four vertices; (b) three subsets with five vertices; (c) three subsets with six vertices; (d) one subset with seven vertices; and (e) one subset with eight vertices.*

Taking into account the complementation, we can formulate Corollary 1.

Corollary 1. *In $\mathbb{Z}^3$, there exist (up to isometry): (b') three subsets with three vertices; (c') three subsets with two vertices; (d') one subset with one vertex; and (e') one subset with zero vertices.*

Remark 3. Let us observe that the twenty-two subsets obtained by using Algorithm 3.1 for $n = 3$ coincide (up to rotations of the 3–dimensional space) with the twenty-two types of unit cell presented by Kong and Roscoe in Figure 1 in [6].

Theorem 2. *In $\mathbb{Z}^4$, there exist (up to isometry): (a) seventy-four subsets with eight vertices; (b) fifty-six subsets with nine vertices; (c) fifty subsets with ten vertices; (d) twenty-seven subsets with eleven vertices; (e) nineteen subsets with twelve vertices; (f) six subsets with thirteen vertices; (g) four subsets with fourteen vertices; (h) one subset with fifteen vertices; and (i) one subset with sixteen vertices.*

Taking into account the complementation, we can formulate Corollary 2.

Corollary 2. *In $\mathbb{Z}^4$, there exist (up to isometry): (b') fifty-six subsets with seven vertices; (c') fifty subsets with six vertices; (d') twenty-seven subsets with five vertices; (e') nineteen subsets with four vertices; (f') six subsets with three vertices; (g') four subsets with two vertices; (h') one subset with one vertex; and (i') one subset with zero vertices.*

Remark 4. The results shown in Theorem 2 and Corollary 2 confirm Pólya's count in 1940 (see Table II in [9]), whose main difficulty to count the different 2–colorings of the 4–dimensional hypercube was the derivation of the appropriate cycle indices (see [2] for more details).

Algorithm 3.1 is based on graph isomorphisms, which is a problem in NP (see [1,4] for more details). For this reason, a more efficient algorithm to ignore the subsets of points that differ by isometries of the n–dimensional space has been implemented. This new algorithm computes the group of isometries of the n–dimensional unit hypercube in $\mathbb{R}^n$ and uses it to ignore the subsets of points of it that differ by isometries of the n–dimensional space. Proposition 7 in [13] proves that an algorithm of this type returns the subsets of points non-congruent that differ by isometries of the n–dimensional space. A scheme of this algorithm is shown in Figure 3.

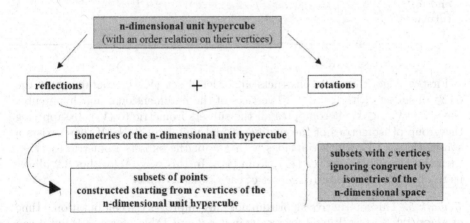

Fig. 3. Scheme of a more efficient algorithm than Algorithm 3.1 to ignore the subsets of vertices of the n–dimensional unit hypercube that differ by isometries of the n–dimensional space

Algorithm 3.2 ignores the subsets of points that differ by isometries from the group *iso_ncube* of isometries of the n–dimensional unit hypercube.

Algorithm 3.2

Input: $(V, \prec)$ set of vertices of the n–dimensional unit hypercube arranged on the order relation $\prec$.

      iso_ncube: group of isometries of the n–dimensional unit hypercube.
// *NIS*: empty list to save the non-isometric subsets with $0 \leq c \leq 2^n$ points.
Output: non-isometric with $0 \leq c \leq 2^n$ points.
begin
  for $c = 0, ..., 2^n$ **do**
      Construct an ordered set $(V_c, \prec)$ containing to the $C(2^n, c)$ subsets with c points
      for $(V_c)_i \subset V_c$ **do**
         iso_vci $= \emptyset$
         {Empty list to save the subsets isometric to $(V_c)_i$.}
         for $\sigma \in$ *iso_ncube* **do**
            iso_vci $=$ *iso_vci* $\bigcup \sigma((V_c)_i)$
         end for
         for all $(V_c)_j \subset V_c$ satisfying $(V_c)_j \in$ *iso_vci* **do**
            $(V_c)_i$ and $(V_c)_j$ are isometric
            $V_c = V_c - \{(V_c)_j\}$
         end for
      end for
      $NIS = NIS \bigcup V_c$
  end for
  return NIS
end

Firstly, Algorithm 3.2 constructs an ordered set $(V_c, \prec)$ containing to the $C(2^n, c)$ subsets with $0 \leq c \leq 2^n$ vertices of the n–dimensional unit hypercube. Next, for $(V_c)_i \subset V_c$, it computes all the subsets isometric to $(V_c)_i$ by applying the group of isometries of the n–dimensional unit hypercube. If there exists a subset $(V_c)_j \subset V_c$ which coincides with any of the subsets isometric to $(V_c)_i$, then the algorithm removes $(V_c)_j$ from (V_c). In this way, Algorithm 3.2 allows us to obtain a representative subset of each isometry class.

Remark 5. This alternative algorithm not only improves the computational time of Algorithm 3.1 (see Table 1 for results in $n = 3$ and Table 2 for $n = 4$) but it can be used for constructing non-isometric subsets of vertices of the n–dimensional unit hypercube, regardless of its cardinality.

Remark 6. In the same way as Algorithm 3.1, the new algorithm determines the smallest non-congruent subsets with respect to the order relation given on the set of vertices of the n–dimensional unit hypercube; so that, by changing the order relation, subsets congruent with these ones are obtained.

Remark 7. Theorems 1 and 2, and Corollaries 1 and 2 are held by using the algorithm whose scheme is shown in Figure 3.

Table 1. Algorithm 3.1 constructs and checks $C(8,4) = 70$, $C(8,5) = 56$, $C(8,6) = 28$ and $C(8,7) = 8$ multi-graphs in 3.12, 3.83, 2.8 and 1.58 seconds of CPU time; respectively. These results are improved to 0.12, (practically) 0, (practically) 0 and (practically) 0 seconds of CPU time; respectively.

	Algorithm 3.1	Alternative algorithm
$c = 4$	3.12 sec.	0.12 sec.
$c = 5$	3.83 sec.	–
$c = 6$	2.8 sec.	–
$c = 7$	1.58 sec.	–

Table 2. Algorithm 3.1 constructs and checks $C(16,8) = 12870$, $C(16,9) = 11440$, $C(16,10) = 8008$, $C(16,11) = 4368$, $C(16,12) = 1820$, $C(16,13) = 560$, $C(16,14) = 120$ and $C(16,15) = 16$ multi-graphs in 11092.3, 9482.27, 11652.5, 8382.22, 2384.01, 734.34, 175.04 and 25.786 seconds of CPU time; respectively. These results are improved to 18.8, 18.56, 14.63, 10.23, 5.43, 2.76, 1.55 and 1.73 seconds of CPU time; respectively.

	Algorithm 3.1	Alternative algorithm
$c = 8$	11092.3 sec.	18.8 sec.
$c = 9$	9482.27 sec.	18.56 sec.
$c = 10$	11652.5 sec.	14.63 sec.
$c = 11$	8382.22 sec.	10.23 sec.
$c = 12$	2384.01 sec.	5.43 sec.
$c = 13$	734.34 sec.	2.76 sec.
$c = 14$	175.04 sec.	1.55 sec.
$c = 15$	25.78 sec.	1.73 sec.

4   Conclusion and Examples

In this paper, we have shown an algorithm more efficient than the extension of that in [8] to compute subsets of points non-congruent by isometries of the n–dimensional space. By using this algorithm for $n = 3$ and $n = 4$, 22 and 402 non-isometric subsets (see Figures 4 and 5 for some examples of these subsets), respectively, have been computed by using a low CPU time. This algorithm allows us to reconstruct objects from the n–xels of n–dimensional binary digital images (see Figure 6 for 2 and 3–dimensional examples).

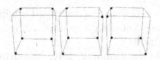

Fig. 4. Non-isometric subsets with five points of $\mathbb{Z}^3$ computed by using Algorithm 3.1 and its alternative using 3.83 and 0 seconds of CPU time, respectively

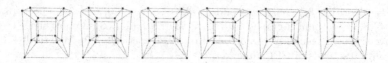

Fig. 5. Non-isometric subsets with thirteen points of $\mathbb{Z}^4$ computed by using Algorithm 3.1 and its alternative using 734.34 and 2.76 seconds of CPU time, respectively

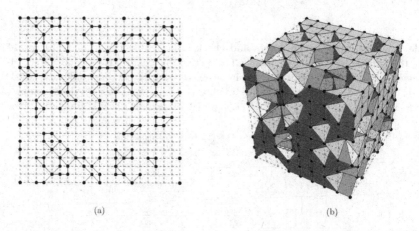

(a)   (b)

Fig. 6. Reconstruction of an object from (a) the 172 pixels (points) of the digital image localized on a grid of size 20×20, (b) the 495 voxels (points) of the digital image localized on a grid of size $8 \times 8 \times 8$

Remark 8. Let us note that the pictures in Figure 6 represent a 2–dimensional (resp. 3–dimensional) object by extracting the boundary of the cell complex whose cells are constructed from the convex hull of the black vertices of each square (resp. cube). Moreover, this technique to represent objects from the boundary of cell complexes can be extended to higher dimensions.

References

1. Arias-Fisteus, J., Fernández-García, N., Sánchez-Fernández, L., Delgado-Kloos, C.: Hashing and canonicalizing Notation 3 graphs. Journal of Computer and System Sciences 76(7), 663–685 (2010)
2. Banks, D.C., Linton, S.A., Stockmeyer, P.K.: Counting Cases in Substitope Algorithms. IEEE Transactions on Visualization and Computer Graphics 10(4), 371–384 (2004)
3. Dörksen-Reiter, H.: Shape representations of digital sets based on convexity properties. Dissertation, Universität Hamburg (2005)
4. Gross, J.L., Yellen, J.: Handbook of Graph Theory. CRC Press, Boca Raton (2003)
5. Han, S.E.: A generalized digital (k_0, k_1)-homeomorphism. Note di Matematica 22(2), 157–166 (2003)

6. Kong, T.Y., Roscoe, A.W.: A theory of binary digital pictures. Computer Vision, Graphics, and Image Processing 32(2), 221–243 (1985)
7. Kong, T.Y., Roscoe, A.W.: Continuous Analogs of Axiomatized Digital Surfaces. Computer Vision, Graphics, and Image Processing 29(1), 60–86 (1985)
8. Pacheco, A., Mari, J.-L., Real, P.: Obtaining cell complexes associated to four dimensional digital objects. Imagen-a 1(2), 57–64 (2010)
9. Pólya, G.: Sur Les Types Des Propositions Composées. The Journal of Symbolic Logic 5(3), 98–103 (1940)
10. Rodrigues Costa, S., Gerônimo, J.R., Palazzo Jr, R., Carmelo Interlando, J., Muniz Silva Alves, M.: The Symmetry Group of $\mathbb{Z}_q^n$ in the Lee Space and the $\mathbb{Z}_{q^n}$-Linear Codes. In: Mattson, H.F., Mora, T. (eds.) AAECC 1997. LNCS, vol. 1255, pp. 66–77. Springer, Heidelberg (1997)
11. Rosenfeld, A.: Connectivity in Digital Pictures. Journal of The ACM - JACM 17(1), 146–160 (1970)
12. Rosenfeld, A.: Three-dimensional digital topology. Information and Control 50(2), 119–127 (1981)
13. Ziegler, G.M.: Lectures on 0/1-polytopes. Oberwolfach Seminars 29, 1–41 (2000)

Computational Topology in Text Mining

Hubert Wagner, Paweł Dłotko, and Marian Mrozek

Institute of Computer Science Jagiellonian University,
{hubert.wagner,pawel.dlotko,marian.mrozek}@ii.uj.edu.pl

Abstract. In this paper we present our ongoing research on applying computational topology to analysis of structure of similarities within a collection of text documents. Our work is on the fringe between text mining and computational topology, and we describe techniques from each of these disciplines. We transform text documents to the so-called vector space model, which is often used in text mining. This representation is suitable for topological computations. We compute homology, using discrete Morse theory, and persistent homology of the Flag complex built from the point-cloud representing the input data. Since the space is high-dimensional, many difficulties appear. We describe how we tackle these problems and point out what challenges are still to be solved.

Keywords: Computational topology, Computational homology, Flag Complex, Discrete Morse theory, Text mining, Vector space model.

1 Introduction and Existing Work

With the growth of the Internet, efficient and accurate information-retrieval systems are of great importance. Modern search-engines are able to quickly query amounts of data counted in exabytes. Text mining aims at performing more in-depth analysis, revealing some additional knowledge from the data.

Text mining methods often use graph-theoretical approaches [11]. Analysing the connected components of the graph of *similarities* between pairs of documents is a simple example. From a topological perspective, such analysis operates on 1-dimensional *complexes* (only pairs of documents are considered) and gives 0-dimensional topological information.

In general, higher dimensional relationships, i.e. relationships between larger subsets of data, are sometimes used in data-mining. For example, the number of triangles (3-cliques) is an important descriptor of the connectivity of a social or collaborative network [6]. Rather than finding just the *number* of such higher-dimensional elements, we would like to compute their topological structure.

We believe that mining a higher dimensional *topological structure* within a set of text documents can give an important insight into the data. In general, the current state-of-the-art topological methods are incapable of handling large datasets in high dimensions, but efficient methods are being developed [15]. Still, we believe that experimenting with smaller, properly sampled data can give interesting insights. For example, [3] shows that data coming from natural images form a topological Klein bottle.

M. Ferri et al. (Eds.): CTIC 2012, LNCS 7309, pp. 68–78, 2012.

In the ongoing research, done in cooperation with Google, we use the tools of computational topology to robustly analyse and compare text data. The goal is to find meaningful topological patterns. This information can help understand the global structure of the data. In a longer perspective, this knowledge can be used in conjunction with the standard methods, improving the quality of information-retrieval systems. This is a novel direction, as is the application of computational topology in higher dimensions. In this paper we show how we adapt existing topological methods and how we tackle computational difficulties, exploiting certain properties of the data. The main question we seek to answer is whether the current computational topology algorithms are capable of efficiently handling reasonable amounts of text data.

For an introduction to computational topology see [5]. A paper by Carlson [3] is an important work, which shows that analysis of higher-dimensional data can be meaningful. A number of papers dealing with lower-dimensional spaces exist, but these techniques are hard or impossible to generalize to higher dimensions [12]. A recent paper by Zomorodian [15] deals with building Rips complexes of high dimensional data, which is also part of our computations. A PhD thesis of Lewiner [9] describes the usage of discrete Morse theory to compute homology groups.

2   Background

2.1   Vector Space Model

We start with describing a way to map textual data into a representation which allows us to use topological tools. Vector space model is a standard tool in information retrieval and data mining [13]. A *corpus*, i.e. collection of text documents, is mapped into points (or vectors) in $\mathbb{R}^n$. These vectors are the so-called *term-vectors* and each of them represents a single document, as described below. Each dimension in this space corresponds to a single word (or *term*).

With each document in a corpus, we associate a term-vector [13], containing words characteristic of this document. In practice from 10 to 50 words are extracted. While term-vectors do not fully describe the documents, they roughly encapsulate the *topic*. Each term t contained in some document d in corpus D is weighted according to the standard *tf-idf* [13] technique: $w(d,t) = tf(d,t) \cdot idf(t)$, where $tf(d,t)$ is the number of occurrences of word t in document d, and $idf(t) = log \frac{|D|}{|\{d:t\in d\}|}$. Thus, more frequent words in a document are weighted higher but this is offset by the *global* popularity of a given term. By P we denote the array of term-vectors representing all the documents of the corpus D. Each term-vector is associated with a unique integer, which is the *index* of that term-vector in P.

In terms of implementation, each term in the corpus can also be assigned a unique number, which represents the term. This is more efficient than storing multiple copies of the string representations of the terms. Term-vector $d \in P$ is compactly stored as a sparse vector: we explicitly represent only the coordinates

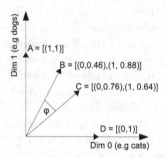

Fig. 1. Example of the vector space model. A two-dimensional space is shown, which means that only two different words are extracted from all documents. The similarity between vectors B, C equals $cos(\varphi) = 0.46 \cdot 0.76 + 0.88 \cdot 0.64 = 0.91$.

with non-zero weights. The actual data-structure representing term-vector d is simply an array of pairs (index of t, $w(d,t)$). See Figure 1 for a simple example. Note that, for brevity, we often identify a document with its term-vector.

Rather than using the Euclidean metric on this space, we use the so-called *cosine similarity measure*. This is a natural choice, as this measure is a standard text mining tool used to compare documents. The similarity between two documents (represented by term-vectors a, b), is given by $sim(a,b) := cos(\angle(a,b)) = \frac{\langle a,b \rangle}{||a|| \, ||b||}$. This formula requires computing square roots, which is costly. We will store normalized (according to Euclidean norm) term-vectors and equivalently compute similarity as:

$$sim(a,b) = \langle a,b \rangle$$

Cosine similarity quantifies the closeness of topics of two documents [13]. The values range from 0 (completely unrelated topics) to 1 (identical topic). Note that the constructed space (equipped with the cosine similarity measure) is not a metric space. Later we will use a weight function $d(a,b) := 1 - sim(a,b)$, which is also not a metric, as the triangle inequality is not satisfied.

We have to distinguish between extrinsic (embedding) and intrinsic dimension of the space. In this case, the extrinsic dimension, R, is large, equal to the number of unique words in the dataset, which can reach tens of thousands to several millions in practical applications. It is typically assumed that the intrinsic dimension is significantly lower, which prevents the *curse of dimensionality* from making the computations infeasible. Another important property of data coming from real-world text corpora is that the frequency of word occurrences follow the Zipf distribution [14]. It assumes that the frequency of an r-th most common word is expressed as: $P(r) = \frac{R}{r\ln(cR)}$, where c is some constant which depends on the corpus. This distribution is far from uniform – intuitively, the most common words apear much more often than the others.

Also, note that due to very high extrinsic dimensionality, the space is very sparse (empty) in practice. Another observation is that the number of keywords extracted from each document is relatively small. In practice, this number is chosen between 10-50. So, for each term-vector, the number of nonzero coordinates is small, compared to the number of zero coordinates. Therefore, the similarity between two randomly chosen term-vectors should be zero most of the time, since the support of these vectors is disjoint. These facts suggest that this space behaves differently than the Euclidean space, where the distance between any two points is finite (intuitively, zero similarity corresponds to infinite distance). In practice, this effect is offset by the Zipf distribution – more popular words increase the number of pairs of documents with nonzero similarity.

The described properties of the data are important, as they reduce the number of large cliques appearing during computations. This makes topological computations based on flag complexes, as described in the following section, more feasible.

2.2 Computational Topology

First, we would like to outline the computations we perform. We are interested in computing *homology* and *persistent homology* of the space describing similarities between the documents in a corpus. Representing the textual data in the vector space model yields a point-cloud, allowing us to use topological tools. Starting from the point-cloud we will construct a *simplicial complex* called a *flag complex*, which encodes higher dimensional topological information, and can be viewed as a higher-dimensional analog of a graph. Since the complex can be large, we simplify it, using discrete Morse theory. This step retains the topological information. Finally, we compute homology on the reduced complex.

A finite collection of sets, S, is an abstract *simplicial complex* if for every $t \in S$ and for every $s \subset t$ we have $s \in S$. Every element $t \in S$ is a *simplex* and its *dimension* is defined as $card(t) - 1$. By S_k we denote the k-skeleton of complex S, i.e. all simplices in S with dimension $\leq k$. If $p \subset q$ and $card(q) - card(p) = 1$, we say that p is a *face* of q and q is a co-face of p. *(Co-)boundary* is the set of all (co-)faces of a simplex. A simplex of dimension 0,1,2 is respectively: *a vertex, an edge* and *a triangle*.

An ϵ-*graph* imposed by the similarity measure sim on the collection of term-vectors P is defined as $G = (P, E)$, where $E = \{(a, b) \in P \times P \mid 1 - sim(a, b) \leq \epsilon\}$. In other words, edges connect pairs of documents with similarity above certain threshold. In general, for graph $G = (V, E)$, a subset $V' \subset V$ is a *clique* if for every $v_1, v_2 \in V'$, $(v_1, v_2) \in E$. *Flag complex* of graph G is defined as: $S(G) := \{V' \subset V \mid V' \text{ is a clique in } G\}$. The flag complex of a graph G is the maximal simplicial complex having G as its 1-skeleton.

3   Construction of Flag Complex

The flag complex, as well as the so-called *Vietoris-Rips* complex (see [5]), is a standard tool used to perform topological data analysis [3]. In this section we describe an efficient bottom-up technique to obtain flag complexes. The presented technique avoids the usage of associative data structures, which incur a significant performance penalty. We designed the code to use only vectors (dynamically growing arrays as in the C++ Standard Library) which are fast due to good caching properties.

The complex building phase is similar to the construction of Vietoris-Rips complex presented in [15]. Since in Section 4 we are focused on computing Morse complexes, we require fast access to the (co-)boundary of each simplex, which is not included in the cited paper. We use the name *flag complex* instead of *Vietoris-Rips complex*, as the latter assumes a metric function.

The input to the algorithms presented in this Section is the array P together with the similarity function *sim*. Let us first describe the data structure we use to store simplices. Each simplex s has a vector `vertices` storing the 0-dimensional simplices (which correspond to indices of term-vectors) that belong to s. Moreover, it has a vector `boundary`, containing the faces of s.

During the construction we also use vectors `coboundary` and `neigh`. Vector s.`neigh` contains the vertices adjacent to *all* vertices in s, with the additional property that for each $i \in s$.`neigh` $i > max\{s.$`vertices`$\}$. We assume the the entire simplex is *created by* its maximum vertex. Importantly, we exploit this property in the algorithms to ensure that each simplex is created only once (when the maximal vertex is processed). `SimplicialComplex` stores a vector of pointers to simplices separately for each dimension. Algorithm 1 shows how we build the 1-skeleton of the constructed flag complex.

Algorithm 1. CreateOneSkeleton

Input: array P of term-vector, double ϵ
1: verts = array of `simplex*`;
2: **for** i = 0 to P.size **do**
3:     verts[i] = new simplex();
4:     verts[i].**vertices** = i;
5: **for** i = 0 to P.size **do**
6:     verts[i].**neigh** = ComputeNeigh(i , P , ϵ);
7: `SimplicialComplex`[0] = verts;
8: edges = array of `simplex*`;
9: **for** for i = 0 to vert.size **do**
10:     **for** j = 0 to vert[i].neigh.size **do**
11:         `simplex*` edge = new `simplex`;
12:         edge.boundary = (vert[i], vert[i].**neigh**[j]);
13:         vert[i].coBoundary.add(edge),   vert[i].**neigh**[j].coBoundary.add(edge);
14:         edge.**vertices** = (vert[i], vert[i].**neigh**[j]);
15:         edge.**neigh** = vert[i].**neigh** $\cap$ vert[i].**neigh**[j].**neigh**;
16:         edges.add(edge);
17: `SimplicialComplex`[1] = edges;

ComputeNeigh algorithm computes the ϵ-neighborhood of a given vertex with a constraint that it returns only vertices with indices higher than the index of the considered vertex. For the time being, we assume that it just iterates through all the vertices, computes the similarity and rejects the vertices corresponding to documents with similarity below the threshold. This makes the complexity of the entire Algorithm 1 quadratic. Since, in practice, the output graph is sparse, the complexity can potentially be reduced. We are investigating methods of computing ϵ-graphs, such as cover trees [2], which are more suitable for this type of data. Some efficient techniques for metric spaces are reviewed in [15].

Once the 1-skeleton of the complex is created, we proceed with the creation of higher dimensional simplexes, as described in Algorithm 2. In terms of computational complexity, the entire constructed flag complex can be exponential in the number of vertices of the input. This is related to the fact, that the total number of cliques is pessimistically exponential. In practice, we are interested in computing the complex only up to a certain, small dimension, which yields polynomial worst-case complexity. The actual performance is heavily dependent on the data.

Algorithm 2. CreateHigherDimensionalSimplices

Input: array initial of **simplex***, int dim

```
 1: new_elements = array of simplex*;
 2: for i = 0 to initial.size do
 3:    for j = 0 to initial[i].neigh.size do
 4:       simplex* new_simplex = new simplex();
 5:       new_simplex.neigh = initial[i].neigh ∩ initial[i].neigh[j].neigh;
 6:       new_simplex.vertices = initial[i].vertices ∪ initial[i].neigh[j]
 7:       initial[i].coboundary.add(new_simplex);
 8:       new_simplex.boundary.add(initial[i]);
 9:       for each bd ∈ initial[i].boundary do
10:          for each cbd ∈ bd.coboundary do
11:             if cbd ≠ initial[i] and initial[i].neigh[j] ∈ cbd.vertices then
12:                cbd.coboundary.add(new_simplex);
13:                new_simplex.boundary.add(cbd);
14:       new_elements.add(new_simplex);
15: SimplicialComplex[dim] = new_elements;
```

---

4   Morse-Flag Complexes

In this section we show how to use discrete Morse theory [7] to compress the Flag complex during its construction. In particular we introduce a novel algorithm which iterates Morse complex computations (see Algorithm 3), yielding the nonpersistent $\mathbb{Z}_2$ homology of the input complex (see Theorem 1). Exploiting this property, we do not have to generate boundary matrices required for algebraic computations, which tend to be costly in terms memory and time. To the best of our knowledge this is the first algorithm to compute homology in general dimension, which does not rely on matrix reduction. The obvious limitation is the fact that is requires field coefficients.

To algorithm proceeds as follows: it computes a Morse matching, then computes the boundaries between the critical (unpaired) cells, which form a Morse complex. Now we repeat the process as long as there exist some nonzero boundaries, which allow to perform the Morse matching.

Note that using $\mathbb{Z}_2$, rather than $\mathbb{Z}$, coefficients simplifies the computations, but prevents us from capturing the so-called *torsion* (see [5]) in homology groups. In terms of implementation, during the construction in dimension n we need to store only the $n-1$ and $n-2$ dimensional elements, which enables us to reduce the memory usage.

Theorem 1. *Let S be the input complex. We build a Morse complex of S and iterate Morse construction, as long as some Morse pairings exist. Let $|S_n|$ denote the number of n-cells in the final Morse complex. Then $\dim H_n(S, \mathbb{Z}_2) = \beta_n(S) = |S_n|$.*

Proof. Since, in general, the Morse complex obtained in the described construction is not a simplicial complex, we use a more general algebraic Morse theory (see Section 11.3, [8]). Compared to the setting of Forman [7], the important difference is that during the matching between cells a and b, where a is a face of b, we want the incidence coefficient ($\kappa(a, b)$, see [8]) to be invertible. However, in case of field coefficients, for a being a face of b we have $\kappa(a, b) \neq 0$, so it is invertible in the field.

We now show that iterative construction of Morse complexes terminates with complex whose all cells have empty boundaries (and consequently coboundaries). Let us assume the contrary, namely that the construction terminated, but some cell a has element b in its boundary. But then the Morse pairing between a and b could have been made, which gives a contradiction.

Let us now assume that we obtained a Morse complex with each cell having empty boundary and coboundary. From algebraic Morse theory we know that its homology is isomorphic to the homology of the initial complex. Homology is defined as the quotient $H_n := ker\partial_n / im\partial_{n+1}$. Since all the boundary maps ∂_n are zero, the image is empty and the kernel group is generated by all the cells. As a result, every cell generates a homology class and consequently $\beta_n(S) = |S_n|$.   □

It might seem that this algorithm contradicts some known hardness results related to constructing optimal gradient vector fields (Morse matchings) [10]. Indeed constructing such a matching on the input complex would imply that homology can be read from the related Morse complex. But we only define the optimal matching on the penultimate complex, and this cannot be transferred back to the original complex.

We will not discuss the complexity of this algorithm. Let us just state that using efficient algorithms to compute Morse complexes from [9,16] we get a bound of $O(n^3)$, where n is the size of the input complex. Experiments show that in practice the runtime is better.

In Algorithm 3 we describe the compression of the initial complex to a Morse complex. The procedure $doMorsePairings(C, d)$ performs Morse pairings in a complex C under a constraint that the dimension of every element in a pairing is $\leq d$. The procedure $computeMorseComplex(C, V)$ computes the boundary coefficients between critical cells (see [7,8] for the theory and also [9] for algorithmic details) and removes from C all the non-critical elements (i.e. elements from V). The procedure stops execution when there are no more pairings to be done in the complex. We want to point out that only the $d-2$-dimensional skeleton of the constructed complex is modified, because we need the simplices in dimension $d-1$ and d in order to build the higher dimensional skeleton. This algorithm should be called for each step of construction in Algorithm 2.

Algorithm 3. IteratedComputationOfMorseComplex

Input: C - initial complex
Output: C - reduced Morse complex
 1: dim = dimension of C;
 2: **while** true **do**
 3:     $n = size(C)$;
 4:     List of Morse pairings V = doMorsePairings(C, $dim-2$);
 5:     computeMorseComplex(C, V);
 6:     **if** $size(C) == n$ **then**
 7:         return C;

---

5   Experiments

We have developed and tested a C++ implementation which includes the algorithms outlined above. To compute homology and persistent homology, we use the standard *matrix reduction* method [5], with the *twist* by Chen and Kerber [4]. For experiments we sample the corpus of the English Wikipedia [17], processed using Python library gensim [18].

There are two main parameters of our software. Parameter dim controls the maximum dimension of the constructed complex. As a result, homology is computed up to dimension $dim-1$. The second parameter is $\epsilon \in [0,1]$, which means that only edges (a,b) with $sim(a,b) \geq 1-\epsilon$ are included in the skeleton. These parameters reduce the amounts of computations.

While in the worst case computing persistent homology takes cubic time, the reduction algorithm is typically assumed to take linear time in practical situations [4]. Judging from our experiments, the behavior is definitely superlinear, probably roughly quadratic (in the size of the complex) for dimensions ≥ 3 (see Figure 2:left). For dimension < 2 the time required to build the complex dominated over the persistence computations. Therefore, for higher dimensions the reduction algorithm is clearly the computational bottleneck. Additionally the number of cells grows super-linearly in the input size, but it is strongly dependent on the chosen ϵ.

The observed quadratic behavior is important in the context of finding efficient methods of computing persistent homology. Recent research suggests that

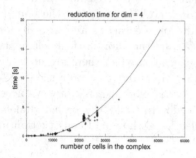
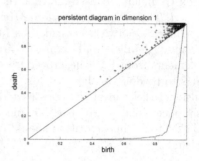

Fig. 2. Left: Runtime (in seconds) of the reduction algorithm for dimension 4. The behaviour appears to be quadratic, which is emphasised by fitting a degree two polynomial. **Right**: Persistence diagram for a complex containing 0.8M cells. We can assume 1 for infinity, since we know that at this point the skeleton would become a complete graph with all the cliques present. The curve in the lower part of the diagram represents a normalized cumulative sum of persistence values. It helps visualize the region of values for which features of relatively large persistence appear.

simplifying the input complex using discrete Morse theory (in a way which retains *persistence*) can increase efficiency. While in practice the significant advantage is in terms of memory usage [16], recent experiments show that such a simplification can minimize the problem related to the slow matrix reduction computation.

The observed quadratic behaviour is a good motivation for further development of algorithms based on discrete Morse theory. In case of textual data, such methods could help increase both the memory and time efficiency of persistence computations. Unfortunately, currently published methods, giving optimality in terms of the number of critical cells extracted, are limited to dimension ≤ 3.

Our attempt to compute *standard* homology using discrete Morse theory was not very effective as the reduction factor was only about $10 - 20\%$. This is probably related to the fact that we have many cycles in the highest dimension. We plan to investigate this issue further.

As shown in Figure 2:right, the 1-dimensional topology is quite uninteresting until the filtration value around 0.8. It means that only after introducing edges with similarity ≤ 0.2, do many features of larger persistence start to appear. On the other hand, we measured that the highest-dimensional cells are the most abundant in the complex. It appears that cells rarely cluster to create homological features of non-zero persistence. Experiments in higher dimensions (but for much smaller datasets) back up this statement. Note that in our setting a boundary of a single simplex either remains a cycle 'forever' (which actually means it is killed at value 1, so persistence equals $1-$birth) or is filled instantly (zero persistence).

These observations suggest that features of non-zero persistence capture, let us call it, semi-similar sets of documents. By that we mean sets of documents which are similar enough to create a, say, p-dimensional cycle, but for a given threshold, they cannot fill this cycle. But, for a higher threshold, there can exist additional cells (related to additional documents) which do fill the cycle. Therefore, persistence can be viewed as the measure of discrepancy between the inter-similarity of a set of documents, and its certain superset.

Analysis of the 1-dimensional persistence suggests two explanations of the low number of features of non-zero persistence. 1) The similarities are very strong, many large cliques appear and most lower-dimensional features have zero persistence. 2) The similarities are strong only locally - almost all appearing cycles are boundaries of a single simplex, so their persistence is 0 (if they are killed). To verify this hypothesis experiments in higher (> 4) dimensions should be run.

6 Summary

The main purpose of this paper is to challenge the current computational topology tools with large text datasets represented within the vector-space model. The experimental results show that these methods lack efficiency. Specifically, the algorithm for persistent homology exhibits quadratic behaviour in the size of the constructed complex, which prevents our approach from scaling for realistic amounts of data. Interestingly, to the best of our knowledge, this is the first dataset exhibiting quadratic scaling, which comes from an application.

The experiments we were able to conduct did shed some light on the lower-dimensional topological structure of the dataset. In our future research we would like to answer some of the questions posed in the previous section as well as try to verify the efficiency of different computational methods.

Acknowledgments. All authors are supported by Google Research Awards programme. H.W. is also supported by a Foundation for Polish Science IPP Programme "Geometry and Topology in Physical Models". M.M. is partially supported by Polish MNSzW Grant N N201 419639, P.D. also by Grant Nr IP 2010 046370.

References

1. Baeza-Yates, R., Ribeiro-Neto, B.: Modern information retrieval, p. 192. Addison-Wesley Longman, Reading (1999)
2. Beygelzimer, A., Kakade, S., Langford, J.: Cover trees for nearest neighbor. In: Proc. of ICML 2006 (2006)
3. Carlson, G.: Topology and Data. Bulletin of the AMS 46(2), 255–308 (2009)
4. Chen, C., Kerber, M.: Persistent homology computation with a twist. In; 27th European Workshop on Computational Geometry, EuroCG 2011 (2011)
5. Edelsbrunner, H., Harer, J.L.: Computational Topology. An Introduction. Amer. Math. Soc., Providence (2010)
6. Feng, A.-X., Fu, C.-H., Xu, X.-L., Liu, A.-F., Chang, H., He, D.-R., Feng, G.-L.: An Empirical Investigation on Important Subgraphs in Cooperation-Competition networks. Science (2011)

7. Forman, R.: A User's Guide To Discrete Morse Theory. Séminaire Lotharingien de Combinatoire B48c, 1–35 (2002)
8. Kozlov, D.: Combinatorial Algebraic Topology. Springer (2007)
9. Lewiner, T.: Geometric discrete Morse complexes, PhD Thesis (2005)
10. Lewiner, T., Lopes, H., Tavares, G.: Toward Optimality in Discrete Morse Theory. Experiment. Math. 12(3), 271–286 (2003)
11. Polanco, X., Juan, E.S.: Text Data Network Analysis Using Graph Approach. In: Proc. of InSciT, pp. 586–592 (2006)
12. Robins, V., Wood, P.J., Sheppard, A.P.: Theory and Algorithms for Constructing Discrete Morse Complexes from Grayscale Digital Images. IEEE Trans. Pattern Anal. Mach. Intell. 33, 1646–1658 (2011)
13. Salton, G., Wong, A., Yang, C.S.: A vector space model for automatic indexing. Commun. ACM 18(11), 613–620 (1975)
14. Zipf, G.K.: Human Behavior and the Principle of Least Effort. Addison-Wesley, Cambridge (1949)
15. Zomorodian, A.: Fast construction of the Vietoris-Rips complex. Computers & Graphics 34(3), 263–271 (2010)
16. Günther, D., Reininghaus, J., Wagner, H., Hotz, I.: Memory Efficient Computation of Persistent Homology for 3D Image Data using Discrete Morse Theory. In: Sibgrapi 2011, Maceio, Brazil (2011)
17. English Wikipedia corpus, http://dumps.wikimedia.org/enwiki/
18. Gensim Library, http://radimrehurek.com/gensim/

Concentrated Curvature for Mean Curvature Estimation in Triangulated Surfaces

Mohammed Mostefa Mesmoudi, Leila De Floriani, and Paola Magillo

Department of Computer Science, University of Genova,
Via Dodecaneso 35, 16146 Genova, Italy
mmesmoudi@ac-creteil.fr, {deflo,magillo}@disi.unige.it

Abstract. We present a mathematical result that allows computing the discrete mean curvature of a polygonal surface from the so-called concentrated curvature generally used for Gaussian curvature estimation. Our result adds important value to concentrated curvature as a geometric and metric tool to study accurately the morphology of a surface.

Keywords: Curvature, Gaussian and mean curvature, Discrete curvature, Triangulated surfaces.

1 Introduction

Curvature is an important geometric tool generally used to study the metric and topological properties of a surface. Indeed, Gauss-Bonnet theorem [10] links the topology of the surface (or of a patch of it) to its total Gaussian curvature. The convexity and concavity of a surface can be studied through mean and Gaussian curvatures and its main morphological features can be deduced from the the critical values of mean curvature. The behavior of geodesic segments (i.e. the shortest segment linking two points on a surface) can be studied through curvature values and their sign over the surface. Curvature has been widely studied in the smooth case and later in the discrete one where several attempts have been made to give adequate definitions for both Gaussian and mean curvatures. Discrete methods either interpolate the discrete values of the surface by a smooth function, or define discrete approaches that guarantee similar properties as the ones available in the smooth case (see [5] for more details). Such methods are based on approximations and, thus, the values they produce suffer from error optimization and control or from the approximation convergence when refining the mesh to tend to a smooth surface.

Concentrated curvature has been defined by Aleksandrov in [3] as the total curvature of spherical caps that approximate a triangulated surface at its vertices. It turns out that concentrated curvature depends only on the total angle around a vertex and does not depend on the radii of the approximating caps. Concentrated curvature produces, thus, an accurate value for each vertex of the surface and does not suffer from computation errors and convergence problems (see Section 3). Moreover, concentrated curvature satisfies a discrete version of

M. Ferri et al. (Eds.): CTIC 2012, LNCS 7309, pp. 79–87, 2012.

the above mentioned Gauss-Bonnet theorem that links the topology of a surface to its metric [10].

In [7], we have introduced *discrete distortion* as a generalization of concentrated curvature to three-combinatorial manifolds, and in [8], we have shown that its restriction to surface boundary of volumetric shape gives a good discrete estimator of mean curvature.

The aim of this paper is to show that concentrated curvature is linked to the restriction of discrete distortion via a simple relation that makes the computation of mean curvature possible from concentrated curvature. As a consequence, principal curvature computation becomes possible as the solution of two simple equations. This result gives to concentrated curvature a crucial role in combinatorial geometry to study the metric properties of a surface.

The reminder of this paper is organized as follows. In Section 2, we present some theoretical background on analytic curvatures. In Section 3, we present concentrated curvature as a Gaussian curvature estimator. In Section 4, we describe how concentrated curvature can be generalized to 3-dimensional manifolds and how its restriction to the boundary surfaces defines a new mean curvature estimator, called discrete distortion. In Section 5, we present the duality between concentrated curvature and discrete distortion. Finally, in Section 6, we present some experiments that highlight such duality, and we draw some conclusions and directions of future development.

2   Background Notions

In this section, we briefly review some fundamental notions on curvature (see [4] for details). Let C be a curve having parametric representation $(c(t))_{t \in R}$. The curvature $k(p)$ of C at a point $p = c(t)$ is given by

$$k(p) = \frac{1}{\rho} = \frac{|c'(t) \wedge c''(t)|}{|c'(t)|^3},$$

where ρ, called the *curvature radius*, corresponds to the radius of the osculatory circle tangent to C at p.

Let S be a smooth surface (at least C^2). Let $\vec{n_p}$ be the normal vector to the surface at a point p. Let Π be the plane which contains the normal vector $\vec{n_p}$. Plane Π intersects S at a curve C containing p: the curvature k_p of C at point p is called *normal curvature* at p. When plane Π turns around $\vec{n_p}$, curve C varies. There are two extremal curvature values $k_1(p) \leq k_2(p)$ which bound the curvature values of all curves C. The corresponding curves C_1 and C_2 are orthogonal at point p [4]. These extremal curvatures are called *principal normal curvatures*. Since the surface is smooth, then *Euler formula* (also called *Dupin indicatrix*) indicates that the curvatures at a point p have an elliptic behavior described by $k(p) = k_1(p)\cos^2(\theta) + k_2(p)\sin^2(\theta)$, where parameter $\theta \in [0; 2\pi]$. The *Gaussian curvature* $K(p)$ and the *mean curvature* $H(p)$ at point p are the quantities

$$K(p) = k_1(p) * k_2(p), \qquad (1)$$

and

$$H(p) = \frac{1}{2\pi} \int_0^{2\pi} k(p)d\theta = \frac{k_1(p) + k_2(p)}{2}.$$ (2)

Gaussian curvature and the mean curvature strongly depend on the (local) geometrical shape of the surface. *Mean curvature* can identify saddle regions and ridge/ravine lines, and mean curvature combined with *Gaussian curvature* can identify convex, concave and locally flat regions. These are relevant properties of curvature for surface analysis:

- Let p be a point with positive Gaussian curvature (i.e., both principal curvatures have the same sign). If the mean curvature is positive [negative] at p, then the surface is locally convex [concave] at p.
- A negative Gaussian curvature at a point p implies that the principal curves lie in two different half spaces with respect to the tangent plane, and thus p is *a saddle point*.
- If the principal curvatures at a point p are null (i.e., the Gaussian and the mean curvatures are null), then the surface is "infinitesimally" *flat* at p.
- If the Gaussian curvature is null and the mean curvature is different from zero at a point p, then the surface is flat in one principal direction and convex [concave] in the other one (if the mean curvature of p is positive or negative, respectively). *Ridge and ravine* lines correspond to such a situation.

A remarkable property of Gaussian curvature is given by *Gauss-Bonnet Theorem*, which relates the metric property given by the Gaussian curvature to the topology of the surface (given by its Euler characteristic) [4].

Theorem 1 (Gauss-Bonnet Theorem). *For a compact surface S with a possible boundary components ∂S we have*

$$\int\int_S K(p)ds + \int_{\partial S} k_g(p)dl = 2\pi\chi(S),$$ (3)

where χ is Euler characteristic of surface S (i.e., $\chi = 2(1-g)$, where g is the genus of the surface), and k_g denotes the geodesic curvature at boundary points (i.e., the geodesic curvature is the norm of the projection of the normal vector of the curve on the tangent plane to the surface).

3   Concentrated Curvature

In [3] a mathematical definition of a discrete Gaussian curvature has been given by means of angle deflection. The author calls it *concentrated curvature* and justifies mathematically this name. Much more recently in [1,2], other authors propose to use concentrated curvature to define a stable alternative to Gaussian curvature.

Let Σ be a (piecewise linear) triangulated surface and let p be a vertex of the triangle mesh. Let $\Delta_1, \cdots, \Delta_n$ be the triangles incident at p such that Δ_i

and Δ_{i+1} are edge-adjacent. If a_i, b_i are the vertices of triangle Δ_i different from p, then the total angle Θ_p at p, also called conical angle, is given by $\Theta_p = \sum_{i=1}^{n} \widehat{a_i p b_i}$.

Around p the surface is isometric to a cone of angle Θ_p at its apex. If $\Theta_p < 2\pi$, then we can approximate the cone by a spherical cap from its interior. Each point on the cap has a constant Gaussian curvature equal to the square of the inverse of the cap radius. The total Gaussian curvature of the cap is then equal to its area normalized by the radius square. By simple computation, this number is equal to $2\pi - \Theta_p$ and is radius independent. This fact implies that approximating the cone by smaller caps, the total Gaussian curvature is always the same. This leads us to the definition of concentrated curvature.

Definition 1. *[10] The* concentrated Gaussian curvature $K_C(p)$, *at a vertex* p *of the triangulated surface, is the value*

$$K_C(p) = \begin{cases} 2\pi - \Theta_p \text{ if } p \text{ is an interior vertex, and} \\ \pi - \Theta_p \;\; \text{if } p \text{ is a boundary vertex,} \end{cases}$$

where Θ_p is the conical angle at p.

For an internal vertex, the quantity $2\pi - \Theta_p$ is computed by approximating the surface at each vertex by spherical caps. The total curvature of each spherical cap is equal to $2\pi - \Theta_p$ and does not depend on the radius of the cap. The detailed justification can be found in [6].

Thus, concentrated curvature is, simply, the angle defect between the flat Euclidean case (i.e., a plane) and the surface. Concentrated curvature for boundary vertices is the angle defect between the case of boundary points of a half plane and the surface.

A simple computation on the number of triangles, edges and vertices within the surface gives the following discrete version of Gauss-Bonnet theorem [10]:

Theorem 2. *Let Σ be a closed orientable triangulated surface, and $\chi(\Sigma)$ be the Euler characteristic of Σ. Then*

$$\sum_{p \text{ vertex of } \Sigma} K_C(p) = 2\pi\chi(\Sigma).$$

4   Discrete Distortion

The principle underlying concentrated curvature can be extended to combinatorial (triangulated) 3-manifolds, by comparing the total solid angle around a vertex with 4π which is the total solid angle around a point in R^3. Let p be a vertex of a combinatorial 3-manifold Ω. *Vertex distortion* at p is thus defined as

$$D(p) = \begin{cases} 4\pi - S_p \text{ if } p \text{ is an interior vertex, and} \\ 2\pi - S_p \text{ if } p \text{ is a boundary vertex,} \end{cases} \tag{4}$$

where S_p is the solid angle at p within the manifold.

We have proven in [7] that, if Σ is a shape embedded in R^3, then internal vertices have null vertex distortion. This is an important property that we use to define the restriction of distortion on the boundary of the 3-manifold without considering the tetrahedra in its interior.

For triangulated surfaces embedded in R^3, the restriction of discrete distortion to a surface reduces to compare the internal solid angles at vertices with 2π. In this case, distortion at a vertex p can be expressed in a simpler way as

$$D(p) = \sum_{e \in St(p)} (\pi - \Theta_e), \tag{5}$$

where $St(p)$ is the set of edges incident to p, and Θ_e is the dihedral angle around edge e. In [8], we have shown, through the use of Conolly functions, that the restriction of distortion to surfaces provides a good discrete approximation of mean curvature.

Mean curvature of a polyhedral surface is usually defined in literature (see, e.g., [9]) by

$$|H| = \tfrac{1}{4|A|} \sum_{i=1}^{n} \|\vec{e_i}\| |\pi - \Theta_i|, \tag{6}$$

where $|A|$ is the the area of the Voronoi or barycentric region around a vertex p, e_i is one of the n edges incident in p with a dihedral angle Θ_i. Formula (6) produces only positive values. A positive or negative sign is given depending on the angle formed by the surface normal at p with the vector obtained by summing all edges, weighted with $|\pi - \Theta_i|$. However, there is another issue when using Formula (6) for mean curvature estimation: curvature values depend on the length of the edges incident at vertex p, and, thus, are area-dependent.

5 Concentrated Curvature versus Discrete Distortion

We show here that there is a natural duality between discrete distortion and concentrated curvature. Let p be a vertex on a triangulated surface Σ embedded in the Euclidean space. Let $(\Delta_i = u_i p u_{i+1})_{i=1\cdots n}$ be the set of all triangles incident at p on Σ and let $(\vec{N_i})_{i=1\cdots n}$ be their unit normal vectors. Vectors $\vec{N_i}$ generate a polyhedral cone $\mathcal{C}(p)$ of summit p where each face F_i ($i = 1 \cdots n$) is defined by two consecutive vectors $\vec{N_i}$ and $\vec{N_{i+1}}$ ($i = 1 \cdots n \ mod(n)$), see Figure 1. Vertex p belongs thus to two surfaces Σ and $\mathcal{C}(p)$.

The following theorem implies that concentrated curvature can be used in different ways to estimate both Gaussian and mean curvatures through simple geometric constructions.

Theorem 3. *Concentrated curvature and distortion of surfaces Σ and $\mathcal{C}(p)$ at vertex p are linked by the following formulas, where indexes refer to the corresponding surface:*

$$D_{\mathcal{C}}(p) + K_{\Sigma}(p) = 2\pi, \quad and \quad D_{\Sigma}(p) + K_{\mathcal{C}}(p) = 2\pi. \tag{7}$$

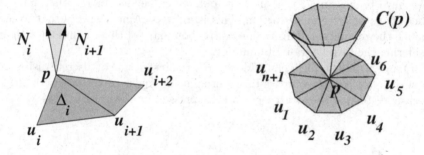

Fig. 1. Duality between distortion and concentrated curvature. Unit normal vectors to triangles incident to p generate a cone $\mathcal{C}(p)$.

Proof. Let $\widehat{u_i}$ be the dihedral angle at edge pu_i shared by triangles Δ_{i-1} and Δ_i. Similarly, let $\widehat{N_i}$ be the dihedral angle at edge $\vec{N_i}$ within the cone $\mathcal{C}(p)$. Simple geometric considerations, imply that the angle between $\vec{N_i}$ and $\vec{N_{i+1}}$ is given by

$$\sphericalangle(\vec{N_i}, \vec{N_{i+1}}) = \pi - \widehat{u_{i+1}}. \tag{8}$$

Conversely, vectors $\vec{pu_i}$ are perpendicular to triangles generated by $(p, \vec{N_{i-1}}, \vec{N_i})$ of cone $\mathcal{C}(p)$. The above relation implies that

$$\widehat{u_{i-1}pu_i} = \sphericalangle(\vec{pu_{i-1}}, \vec{pu_i}) = \pi - \widehat{N_i}. \tag{9}$$

Hence, there is a duality between angles at p of its incident triangles on surface Σ and dihedral ones on cone $\mathcal{C}(p)$, and vice versa. The above results, together with (5), imply that the distortion at p on surface Σ is equal to the total angle at p of all triangles on $\mathcal{C}(p)$, and vice versa. Hence we have:

$$D_{\Sigma}(p) = \sum_{i=1}^{n} \sphericalangle(\vec{N_i}, \vec{N_{i+1}}), \qquad D_{\mathcal{C}}(p) = \sum_{i=1}^{n} (\widehat{u_i pu_{i+1}}). \tag{10}$$

On the other hand, we know that concentrated curvature is the angle deficit on the sum of all triangles incident to a vertex on a surface. Then we have

$$D_{\Sigma}(p) + 2\pi - \sum_{i=1}^{n} \sphericalangle(\vec{N_i}, \vec{N_{i+1}}) = 2\pi, \tag{11}$$

and

$$D_{\mathcal{C}}(p) + 2\pi - \sum_{i=1}^{n} (\widehat{u_i pu_{i+1}}) = 2\pi, \tag{12}$$

which leads to relations (7), and therefore proves the theorem.

Principal curvatures k_1 and k_2 can be obtained as a common solution of both equation $k_1 + k_2 = 2D(p)$ and $k_1 \times k_2 = K(p)$. The result expressed by Theorem 3 provides a new interesting use for concentrated curvature and allows, with the corresponding principal curvatures, a local control of geometry via dual cones, in addition to its topological role described by the discrete Gauss-Bonnet theorem [10].

6   Concluding Remarks

We have implemented the methods defined in (4) and (5) to compute discrete distortion. We have experimentally compared the results and evaluated the efficiency of the computation. In Table 1, we report the order of magnitude of the difference between the values obtained with each method. We can see that the difference in distortion values computed with the two methods is negligible (less than $1/10^9$ of the values range). Moreover, the version with cone angles is slightly faster.

Figures 2 and 3 show the values of distortion and of mean curvature estimated with equation (6), with computation of sign, in a color scale. Color corresponds to negative and to positive values, respectively, in the two figures, and white corresponds to the remaining values. The two methods give the same image.

Table 1. Range of values of distortion, order of magnitude of the maximum difference between the two methods, and execution times (averaged over 100 executions), in seconds

Mesh	Vertices	Distortion range	Difference	Execution time (1)	(2)
Bunny	34k	$[-4.8, 5.2]$	e^{-09}	.398	.378
Bumpy Torus	17k	$[-7.4, 6.4]$	e^{-11}	.192	.185
Octopus	17k	$[-5.6, 6.1]$	e^{-11}	.199	.187
Kitten	11k	$[-6.1, 6.3]$	e^{-11} ·	.128	.122
Happy Buddha	544k	$[-8.9, 2.6]$	e^{-14}	6.43	6.31

We have shown that Gaussian and mean curvatures can be described through a pair of concentrated curvature values at each vertex. This gives concentrated curvature an additional geometric role besides its topological role described by the discrete Gauss-Bonnet theorem.

Surfaces where mean curvature is null everywhere, called minimal surfaces, play a great role in many scientific fields (DNA structures, architecture, etc.). In mathematics, generating and tracking such surfaces is a hard problem due to the complexity of their defining PDE equations. Our work can help studying minimal surfaces through the duality between concentrated curvature and distortion.

Troyanov has shown in [10] that, given a set of points with a corresponding set of weights, then, under some conditions, there exists a polyhedral surface whose vertices are the given set of points and whose concentrated curvatures are the corresponding weights. We project to exploit such result to construct and study the *dual* surface whose vertices are the same as the original surface and whose concentrated curvatures are those data (i.e., concentrated curvature values) coming from the dual cones that we have constructed here. This may reveal other interesting properties linking concentrated curvature to distortion or reveal geometrical and topological properties relating the two surfaces.

Fig. 2. Positive values of distortion (left) and mean curvature (right) in false colors: red (dark grey for black-and-white version) represent positive values, white represents negative or null values

Fig. 3. Negative values of distortion (left) and mean curvature (right) in false colors: blue (dark grey for black-and-white version) represent negative values, white represents positive or null values

Acknowledgements. This work has been partially supported by the National Science Foundation under grant number IIS-1116747, and by the Italian Ministry of Education and Research under the PRIN 2009 program.

References

1. Akleman, E., Chen, J.: Insight for Practical Subdivision Modeling with Discrete Gauss-Bonnet Theorem. In: Kim, M.-S., Shimada, K. (eds.) GMP 2006. LNCS, vol. 4077, pp. 287–298. Springer, Heidelberg (2006)
2. Alboul, L., Echeverria, G., Rodrigues, M.A.: Discrete curvatures and Gauss maps for polyhedral surfaces. In: European Workshop on Computational Geometry (EWCG), Eindhoven, The Netherlands, pp. 69–72 (2005)
3. Aleksandrov, P.: Topologia Combinatoria. Edizioni Scientifiche Einaudi, Torino (1957)
4. Do Carno, M.P.: Differential Geometry of Curves and Surfaces. Prentice-Hall Inc., Englewood Cliffs (1976)
5. Gatzke, T.D., Grimm, C.M.: Estimating curvature on triangular meshes. International Journal on Shape Modeling 12, 1–29 (2006)
6. Mesmoudi, M.M., Danovaro, E., De Floriani, L., Port, U.: Surface segmentation through concentrated curvature. In: Proc. International Conference on Image and Pattern Processing, pp. 671–676, Modena, Italy (2007)
7. Mesmoudi, M.M., De Floriani, L., Port, U.: Discrete distortion in triangulated 3-manifolds. Computer Graphics Forum 27(5), 1333–1340 (2008)
8. Mesmoudi, M.M., De Floriani, L., Magillo, P.: Discrete Distortion for Surface Meshes. In: Foggia, P., Sansone, C., Vento, M. (eds.) ICIAP 2009. LNCS, vol. 5716, pp. 652–661. Springer, Heidelberg (2009)
9. Meyer, M., Desbrun, M., Schroder, M., Barr, A.H.: Discrete differential-geometry operators for triangulated 2-manifolds. In: Hege, H.-C., Polthier, K. (eds.) Proceedings VisMath 2002, Berlin, Germany (2002)
10. Troyanov, M.: Les surfaces Euclidiennes à singularités coniques. L'Enseignement Mathématique 32, 79–94 (1986)

Deletion of (26, 6)-Simple Points
as Multivalued Retractions

Carmen Escribano, Antonio Giraldo, and María Asunción Sastre*

Departamento de Matemática Aplicada, Facultad de Informática
Universidad Politécnica, Campus de Montegancedo
Boadilla del Monte, 28660 Madrid, Spain
{cescribano,agiraldo,masastre}@fi.upm.es

Abstract. In a recent paper we have introduced a notion of multivalued continuity in digital spaces which extends the usual notion of digital continuity and allows to define topological notions, like retractions, in a far more realistic way than by using just single-valued digitally continuous functions. In particular, we have characterized the deletion of simple points in 2-D, one of the most important processing operations in digital topology, as a particular kind of retraction. In this work we extend some of these results to 3-dimensional digital sets.

Keywords: Digital topology, continuous multivalued function, simple point, retraction.

1 Introduction

Digitally continuous maps were first introduced by A. Rosenfeld [15] in 1986. He characterized them as functions taking connected sets to connected sets.

Results related with this type of continuity were proved also by L. Boxer [2–4], who introduced such notions as digital homeomorphism, retracts and homotopies. A different approach using multivalued maps was suggested by V. Kovalevsky [13]. (See the introduction of [5] for a discussion of the limitations of these and related approaches).

In recent papers [5, 6], the authors presented a theory of continuity in digital spaces which extends the one introduced by Rosenfeld. Our approach uses multivalued maps and provides a better framework to define topological notions, like retractions, in a far more realistic way than by using just single-valued digitally continuous functions. This notion has allowed us to characterize most common thinning algorithms for digital images as retractions.

In this work we extend some results of those papers to 3-dimensional digital sets. In sections 2 and 3 we revise the basic notions on digital topology required throughout the paper. In section 4 we recall the notions of a digitally continuous multivalued function and of a digital retraction, formulating one of the results of

---

* The authors have been supported by MICINN MTM2009-07030 (A.Giraldo) and UPM-2011-Q061010133.

M. Ferri et al. (Eds.): CTIC 2012, LNCS 7309, pp. 88–97, 2012.

[5] for 3-dimensional sets (Proposition 2). In section 5 we prove the main result of the paper (Theorem 3): If a point p is (26, 6)-simple in X, then there exists a multivalued $(\mathcal{N}, 26)$-retraction $F : X \longrightarrow X \setminus \{p\}$.

For information on Digital Topology we recommend the survey [11] and the books by Kong and Rosenfeld [12], and by Klette and Rosenfeld [9].

We are grateful to the referees for helpful comments and suggestions.

2 Digital Spaces

Two points in the digital plane $\mathbb{Z}^2$ are 8-adjacent if they are different and their coordinates differ in at most a unit. They are 4-adjacent if they are 8-adjacent and differ in at most a coordinate. Given $p \in \mathbb{Z}^2$ we define $\mathcal{N}(p)$ as the set of points 8-adjacent to p, i.e. $\mathcal{N}(p) = \{p_1, p_2, \ldots, p_8\}$. This is also denoted as $\mathcal{N}_8(p)$. Analogously, $\mathcal{N}_4(p)$ is the set of points 4-adjacent to p (with the above notation $\mathcal{N}_4(p) = \{p_2, p_4, p_6, p_8\}$).

Two points of the digital 3-space $\mathbb{Z}^3$ are 26-adjacent if they are different and their coordinates differ in at most a unit. They are called 18-adjacent if they are 26-adjacent and differ in at most two coordinates, and they are called 6-adjacent if they are 26-adjacent and differ in exactly one coordinate. We have therefore, three different neighborhoods of p: $\mathcal{N}_{26}(p) = \mathcal{N}(p)$, $\mathcal{N}_{18}(p)$ and $\mathcal{N}_6(p)$.

In an analogous way, adjacency relations are defined in $\mathbb{Z}^n$ for $n \geq 4$.

A k-path P in $\mathbb{Z}^n$ from the point q_0 to the point q_r is a sequence $P = \{q_0, q_1, q_2, \ldots, q_r\}$ of points such that q_i is k-adjacent to q_{i+1}, for every $i \in \{0, 1, 2, \ldots, r-1\}$. If $q_0 = q_r$ then it is called a closed path. A set $S \subset \mathbb{Z}^n$ is k-connected if for every pair of points of S there exists a k-path contained in S joining them. A k-connected component of S is a k-connected maximal set.

Given $X \subset \mathbb{Z}^n$, $p \in X$, we say, according to [12], that p is a k-boundary point of X if $\mathcal{N}_{\bar{k}}(p) \cap (\mathbb{Z}^n \setminus X) \neq \emptyset$, where $(k, \bar{k}) = (8, 4)$ if $n = 2$, $(k, \bar{k}) = (26, 6)$ if $n = 3$ (this notation will be used throughout the paper). We denote by $\partial_k X$ the set of k-boundary points of X.

A point p and a set X are k-adjacent if $p \notin X$ and there exists $q \in X$ such that p and q are k-adjacent.

3 Digitally Continuous Single-Valued Functions

We start this section revising the notion of digitally continuous function and some of its properties.

Definition 1. *Let $f : X \subset \mathbb{Z}^m \longrightarrow \mathbb{Z}^n$ be a function between digital spaces with adjacency relations k and k', respectively. According to [15] and [3], f is (k, k')-continuous, if and only if, for every $p, p' \in X$ k-adjacent points of $\mathbb{Z}^m$ then either $f(p) = f(p')$ or $f(p)$ and $f(p')$ are k'-adjacent points of $\mathbb{Z}^n$. When $m = n$ and $k = k'$, f is said to be just k-continuous.*

In [15] several results about digitally continuous functions related to operations with continuous functions, intermediate values property, almost-fixed point theorem, Lipschitz conditions, one-to-oneness, etc, were proved. Boxer [2–4] expanded this notion to digital homeomorphisms, retractions, extensions, homotopies, digital fundamental group, induced homomorphisms, etc. (see also [8] and [10] for previous related results).

In particular, Boxer proved that the k-boundary ∂S of a digital square S is not a digital k-retract of the square [2], i.e., it is not possible to construct a digitally continuous function $f : S \longrightarrow \partial S$ such that $f(x) = x$ for every $p \in \partial S$, as happens if we consider them as subsets of $\mathbb{R}^2$. However, neither the outer k-boundary of an annulus is a k-retract of the annulus, as opposite with what happens considering as subsets of $\mathbb{R}^2$. Therefore, digitally k-continuous single-valued functions can not model correctly the topology of $\mathbb{R}^2$. In the next section we show how it is possible to define a notion of continuity for multivalued functions in such a way that these limitations of digitally continuous single valued functions are alleviated (see Proposition 2).

4   Digitally Continuous Multivalued Functions

The definitions and results in this section were first introduced in [5].

Definition 2. *The r-th subdivision of $\mathbb{Z}^n$ is formed by the set*

$$\mathbb{Z}_r^n := \left\{ \left(\frac{z_1}{3^r}, \frac{z_2}{3^r}, \ldots, \frac{z_n}{3^r} \right) \mid (z_1, z_2, \ldots, z_n) \in \mathbb{Z}^n \right\}$$

and the function $i_r \colon \mathbb{Z}_r^n \longrightarrow \mathbb{Z}^n$ given by $i_r \left(\frac{z_1}{3^r}, \frac{z_2}{3^r}, \ldots, \frac{z_n}{3^r} \right) := (z_1', z_2', \ldots, z_n')$ where $(z_1', z_2', \ldots, z_n')$ is the point in $\mathbb{Z}^n$ which is closest to $\left(\frac{z_1}{3^r}, \frac{z_2}{3^r}, \ldots, \frac{z_n}{3^r} \right)$ in the Euclidean metric.

If we consider in $\mathbb{Z}^n$ a k-adjacency relation, we can consider in $\mathbb{Z}_r^n$ the induced adjacency relation, i.e., $\left(\frac{z_1}{3^r}, \frac{z_2}{3^r}, \ldots, \frac{z_n}{3^r} \right)$ and $\left(\frac{z_1'}{3^r}, \frac{z_2'}{3^r}, \ldots, \frac{z_n'}{3^r} \right)$ are k-adjacent if and only if $(z_1, z_2, \ldots, z_n)$ is k-adjacent to $(z_1', z_2', \ldots, z_n')$.

Proposition 1. *i_r is k-continuous as a function between digital spaces.*

Definition 3. *Given $X \subset \mathbb{Z}^n$, the r-th subdivision of X is the set $X_r := i_r^{-1}(X)$.*

Intuitively, if we consider X made of pixels, (respectively, voxels), the r-th subdivision of X consists in replacing each pixel with 9^r pixels (respectively, 27^r voxels) and the function i_r is like an inclusion in the geometric sense.

Definition 4. *Consider $X, Y \subset \mathbb{Z}^n$. A multivalued function $F \colon X \longrightarrow Y$ is a function F such that for every $x \in X$, $F(x)$ is a non-empty subset of Y. A multivalued function $F \colon X \longrightarrow Y$ is said to be a (k, k')-continuous multivalued* *function if there exists a (k, k')-continuous (single-valued) function from X_r to Y for some $r \in \mathbb{N}$ such that $F(x) := \cup_{x' \in i_r^{-1}(x)} f(x')$. We say then that F is induced by f. When $k = k'$, F is said to be just a k-continuous multivalued function.*

The reader is referred to [5] and [6] for properties of digitally continuous multivalued functions. We just note here that any single-valued digitally continuous function is continuous as a multivalued function, and that if F is a (k, k')-continuous multivalued function then F takes k-connected sets to k'-connected sets.

Definition 5. *Let $X \subset \mathbb{Z}^n$ and $Y \subset X$. We say that Y is a multivalued k-retract of X if there exists a k-continuous multivalued function $F \colon X \longrightarrow Y$ (a multivalued k-retraction) such that $F(y) = \{y\}$ if $y \in Y$. If moreover $F(x) \subset \mathcal{N}(x)$ for every $x \in X \setminus Y$, we say that F is a multivalued $(\mathcal{N}, k)$-retraction.*

The following result, which generalizes a 2-dimensional result proved in [5], and that can be proved in a similar way as that result, shows that digitally continuous multivalued functions and, in particular, multivalued k-retractions and multivalued $(\mathcal{N}, k)$-retractions have similar properties as retractions in $\mathbb{R}^3$, in contrast with single-valued digital retractions, which, as noted in the introduction, have serious limitations to replicate the behavior of retractions in $\mathbb{R}^3$.

Proposition 2. *The following holds:*

  i) The k-boundary of a cube X (with Int $X \neq \emptyset$) is not a multivalued k-retract of X.
  ii) The outer k-boundary $\partial_k X$ of a hollow cube X is a multivalued $(\mathcal{N}, k)$-retract of X.

5 Deletion of Simple Points as $(\mathcal{N}, k)$-Retractions

A point $p \in X \subset \mathbb{Z}^2$ is 8-simple in X if its deletion does not change the topology of X in the sense that, when deleting it, the number of components and holes are preserved. A point $p \in X \subset \mathbb{Z}^3$ is 26-simple in X if, after deleting it, the number and location of components, holes (tunnels) and cavities are preserved (see [1, 14] for the formal definition of these intuitive notions, and [7] for an example of why in the 3-dimensional case it is not enough to preserve just the number of holes).

A k-simple point can be locally detected by the following characterization:

Theorem 1 ([1, 14]). *Let $X \subset \mathbb{Z}^2$. A point $p \in X$ is 8-simple if and only if p is an 8-boundary point of X such that the number of 8-connected components of $\mathcal{N}(p) \cap X$ which are 8-adjacent to p is equal to 1.*

Let $X \subset \mathbb{Z}^3$. A point $p \in X$ is 26-simple if and only if the number of 26-connected components of $\mathcal{N}(p) \cap X$ which are 26-adjacent to p is equal to 1, and the number of 6-connected components of $\mathcal{N}_{18}(p) \cap (\mathbb{Z}^3 \setminus X)$ which are 6-adjacent to p is equal to 1.

The following theorem was proved in [5].

Theorem 2. *Consider $X \subset \mathbb{Z}^2$ and $p \in X$. Then p is an 8-simple point if and only if there exists a multivalued $(\mathcal{N}, 8)$-retraction $F : X \longrightarrow X \setminus \{p\}$.*

Theorem 2 also holds for $(k, \bar{k}) = (4, 8)$ with an additional condition on p.

The following result generalizes to $\mathbb{Z}^3$ the "only if" part of Theorem 2.

Theorem 3. *Consider $X \subset \mathbb{Z}^3$ and $p \in X$ a 26-simple point. Then there exists an $(\mathcal{N}, 26)$-retraction $F : X \longrightarrow X \setminus \{p\}$.*

Proof. We show how to define a multivalued $(\mathcal{N}, 26)$-retraction $F : X \longrightarrow X \setminus \{p\}$ according to the number of points of $\mathcal{N}_6(p) \cap X$.

Denote the points in $\mathcal{N}_6(p)$ as follows: n the point above p, s the point below p, w the point left of p, e the point right of p, f the point front of p, b the point behind p. These 6 directions allows us to determine each point of $\mathcal{N}_{26}(p)$.

<u>Case 1</u>: $\mathrm{card}(\mathcal{N}_6(p) \cap X) \geq 3$. If $\mathrm{card}(\mathcal{N}_6(p) \cap X) = 6$, p is not 26-simple. Then $\mathcal{N}(p)$ is 26-simple if and only if p is, up to symmetries, as in Figure 1, where the points not explicitly drawn can be in X or in $\mathbb{Z}^3 \setminus X$.

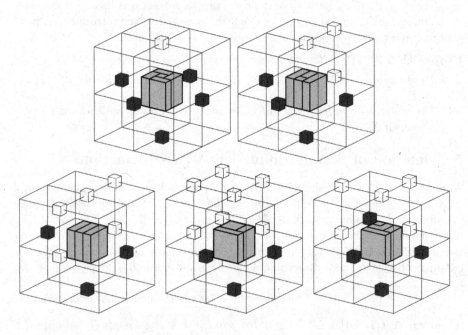

Fig. 1. Local configurations of a 26-simple point with $\mathrm{card}(\mathcal{N}_6(p) \cap X) \geq 3$

If $\mathrm{card}(\mathcal{N}_6(p) \cap X) = 5$, we subdivide p in five parts $(A_1, A_2, A_3, A_4, A_5)$ as in Figure 1a and we define f inducing F such that $f(A_i)$ is the point of $\mathcal{N}_6(p) \cap X$ closest to A_i. Note that f is continuous because if A_i and A_j are 26-adjacent then so are $f(A_i)$ and $f(A_j)$, and if a voxel in $\mathcal{N}(p)$ is 26-adjacent to A_i, then it is also 26-adjacent to $f(A_i)$.

If card$(\mathcal{N}_6(p) \cap X) = 4$, $\mathcal{N}(p)$ is, up to symmetries, as in Figure 1b. Since p is simple, the point nb (north and behind) must be in $\mathbb{Z}^2 \setminus X$ because it is necessary to 6-connect in $\mathcal{N}_{18}(p)$ the points n and b (see Theorem 1). To define F we subdivide p in 4 parts (A_1, A_2, A_3, A_4) and define f inducing F such that $f(A_i)$ is the point of $\mathcal{N}_6(p) \cap X$ closest to A_i.

If card$(\mathcal{N}_6(p) \cap X) = 3$, then p is, up to symmetries, as in one of the configurations of Figures 1c-1e (the configurations correspond to the different ways to 6-connect in $\mathcal{N}_{18}(p)$ the 3 white points of $\mathcal{N}_6(p)$). In all cases we subdivide p as in the figure and we define f inducing F such that $f(A_i)$ is the point of $\mathcal{N}_6(p) \cap X$ closest to A_i.

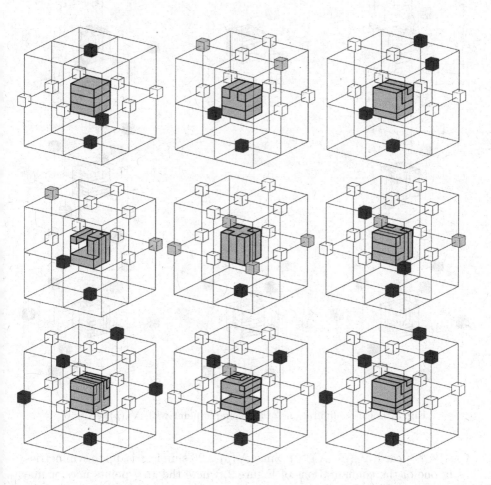

Fig. 2. Local configurations of a 26-simple point with $2 \geq$ card$(\mathcal{N}_6(p) \cap X) \geq 1$. There are three more configurations, interchanging fe and fw in each of the three last configurations (in each case the same subdivisions are used to define F).

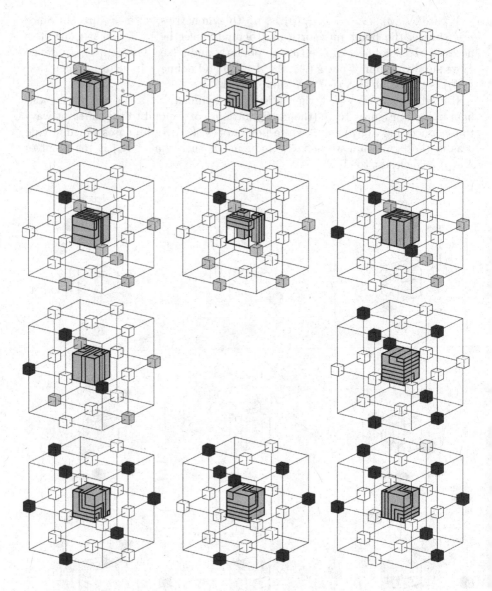

Fig. 3. Local configurations of a 26-simple point with $\mathcal{N}_6(p) \cap X = \emptyset$

<u>Case 2</u>: $2 \geq \operatorname{card}(\mathcal{N}_6(p) \cap X) \geq 1$. Since $\mathcal{N}(p)$ is 26-simple, p is, up to symmetries, as in one of the configurations of Figure 2, where the gray points may or may not be in X. The gray points may depend of their neighbors (for example, in the second case, if nbw is black, then nw must be also black).

In all cases, we subdivide p in up to 5 parts as in Figure 2 (in some cases, to better see the subdivision, one of the subsets is shown transparent) and we define f inducing F such that the image of the smallest parts are always the blue or the gray point they are "oriented to" (i.e. that for which this region is the closest). Each of these parts is adjacent to a bigger part which goes to a blue o gray point adjacent to the former, and so on. If any of the gray points were not in X its part of the subdivision would be absorbed by its neighboring part.

<u>Case 3</u>: $\mathrm{card}(\mathcal{N}_6(p) \cap X) = 0$. If $\mathcal{N}_{18}(p) \cap X = \emptyset$, then $\mathcal{N}(p) \cap X$ consists on just one corner of $\mathcal{N}(p)$, q, and we define $F(p) = q$, In the rest of the cases $\mathcal{N}_{18}(p) \cap X \neq \emptyset$. All possible configurations are shown in Figure 3.

The seven first cases satisfy that there is at least a white point q 6-adjacent to p with $\mathrm{card}(\mathcal{N}_6(q) \cap \mathcal{N}(p) \cap (\mathbb{Z}^3 \setminus X)) \geq 3$. The four last cases satisfy that $\mathrm{card}(\mathcal{N}_6(q) \cap \mathcal{N}(p) \cap (\mathbb{Z}^3 \setminus X)) \leq 2$ for all q 6-adjacent to p.

Note that, as in case 2, the gray points may depend of their neighbors (for example, in the first case, if nbw is black, then bw must be also black). Moreover, not all combinations of gray points are possible (for example, in the second case, if fe is white, all the gray points must be also white).

In all cases, we subdivide p in up to 7 parts as in Figure 3 (in some cases, to better see the subdivision, one of the subsets is shown transparent), and we define f inducing F following the same rule as in case 2 (now, when two points can be chosen as the next one, only one of them allows to define the image of all the remaining subsets).

In Figure 4 we show the steps to construct the subdivision in one of the three last cases (the ninth) which are more involved.

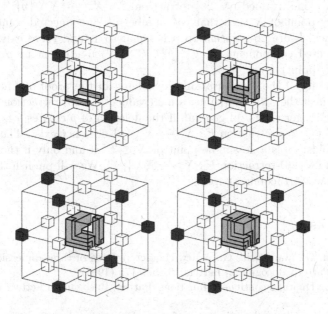

Fig. 4. Construction of the subdivision of p in one case

Remark 1. The previous theorem guarantees that, if $p \in X \subset \mathbb{Z}^3$ is a 26-simple point, then there exists an $(\mathcal{N}, 26)$-retraction $F : X \longrightarrow X \setminus \{p\}$.

Moreover, the multivalued $(\mathcal{N}, 26)$-retraction $F : X \longrightarrow X \setminus \{p\}$ constructed in the proof of the theorem is given by $F(x) = \{x\}$ if $x \neq p$ and

$$F(p) = (\mathcal{N}_6(p) \cap X) \cup \{q \in \mathcal{N}^*_{18}(p) \cap X \mid \mathcal{N}_6(p) \cap \mathcal{N}_6(q) \cap X = \emptyset\}$$
$$\cup \{q \in \mathcal{N}^*_{26}(p) \cap X \mid \mathcal{N}_{18}(p) \cap \mathcal{N}_{18}(q) \cap X = \emptyset\},$$

where $\mathcal{N}^*_{18}(p) = \mathcal{N}_{18}(p) \setminus \mathcal{N}_6(p)$ and $\mathcal{N}^*_{26}(p) = \mathcal{N}_{26}(p) \setminus \mathcal{N}_{18}(p)$. Note that, since p is 26-simple, the third term in the union is nonempty if and only if $\mathcal{N}(p) \cap X$ is just one corner of $\mathcal{N}(p)$.

Analogously, it can be easily seen, using the techniques of [5,6], that if $X \subset \mathbb{Z}^2$ and $p \in X$, then the multivalued function $F : X \longrightarrow X \setminus \{p\}$ given by $F(x) = \{x\}$ if $x \neq p$ and

$$F(p) = (\mathcal{N}_4(p) \cap X) \cup \{q \in \mathcal{N}^*_8(p) \cap X \mid \mathcal{N}_4(p) \cap \mathcal{N}_4(q) \cap X = \emptyset\}.$$

is a multivalued $(\mathcal{N}, 8)$-retraction if and only if p is an 8-simple point.

6   Conclusions and Future Work

We have shown that if p is a 26-simple point of $X \subset \mathbb{Z}^3$ then there exists a multivalued $(\mathcal{N}, 26)$-retraction $F : X \longrightarrow X \setminus \{p\}$.

Conversely, it can be proved, in a similar way as for Theorem 2, that, if there exists a multivalued $(\mathcal{N}, 26)$-retraction $F : X \longrightarrow X \setminus \{p\}$, then p is a 26-boundary point of X such that the number of 26-connected components of $\mathcal{N}(p) \cap X$ which are 26-adjacent to p is equal to 1, and there exists at least one 6-connected components of $\mathcal{N}_{18}(p) \cap (\mathbb{Z}^3 \setminus X)$ 6-adjacent to p (i.e p is a 26-boundary point).

Moreover, in all the configurations considered in the proof of Theorem 3, if we 6-disconnect the "white" points, a multivalued $(\mathcal{N}, 26)$-retraction $F : X \longrightarrow X \setminus \{p\}$ not longer exists. In general, if the deletion of p creates a hole in X, a multivalued $(\mathcal{N}, 26)$-retraction $F : X \longrightarrow X \setminus \{p\}$ does not exist. Therefore, the following holds: "p is a 26-simple point of $X \subset \mathbb{Z}^3$ if and only if there exists a multivalued $(\mathcal{N}, 26)$-retraction $F : X \longrightarrow X \setminus \{p\}$". We will publish the proof of this result in a forthcoming paper.

References

1. Bertrand, G., Mandalain, G.: A new characterization of three-dimensional simple points. Pattern Recognition Letters 15, 169–175 (1994)
2. Boxer, L.: Digitally continuous functions. Pattern Recognition Letters 15, 833–839 (1994)
3. Boxer, L.: A Classical Construction for the Digital Fundamental Group. Journal of Mathematical Imaging and Vision 10, 51–62 (1999)

4. Boxer, L.: Properties of Digital Homotopy. Journal of Mathematical Imaging and Vision 22, 19–26 (2005)
5. Escribano, C., Giraldo, A., Sastre, M.A.: Digitally Continuous Multivalued Functions. In: Coeurjolly, D., Sivignon, I., Tougne, L., Dupont, F. (eds.) DGCI 2008. LNCS, vol. 4992, pp. 81–92. Springer, Heidelberg (2008)
6. Escribano, C., Giraldo, A., Sastre, M.A.: Digitally continuous multivalued functions, morphological operations and thinning algorithms. Journal of Mathematical Imaging and Vision 42(1), 76–91 (2012)
7. Fourey, S., Malgouyres, M.: A concise characterization of 3D simple points. Discrete Applied Mathematics 125, 59–80 (2003)
8. Khalimsky, E.: Topological structures in computer science. Journal of Applied Mathematics and Simulation 1, 25–40 (1987)
9. Klette, R., Rosenfeld, A.: Digital Geometry. Elsevier (2004)
10. Kong, T.Y.: A digital fundamental group. Comp. Graphics 13, 159–166 (1989)
11. Kong, T.Y., Rosenfeld, A.: Digital Topology: Introduction and survey. Computer Vision, Graphic and Image Processing 48, 357–393 (1989)
12. Kong, T.Y., Rosenfeld, A. (eds.): Topological algorithms for digital image processing. Elsevier (1996)
13. Kovalevsky, V.: A new concept for digital geometry. In: Ying-Lie, O., et al. (eds.) Shape in Picture. Proc. of the NATO Advanced Research Workshop, Driebergen, The Netherlands (1992); Computer and Systems Sciences 126. Springer (1994)
14. Rosenfeld, A.: Digital topology. The American Mathematical Monthly 86(8), 621–630 (1979)
15. Rosenfeld, A.: Continuous functions in digital pictures. Pattern Recognition Letters 4, 177–184 (1986)

Topological Operators on Cell Complexes in Arbitrary Dimensions

Lidija Čomić[1] and Leila De Floriani[2]

[1] University of Novi Sad
comic@uns.ac.rs
[2] University of Genova
deflo@disi.unige.it

Abstract. Cell complexes have extensively been used as a compact representation of both the geometry and topology of shapes. They have been the basis modeling tool for boundary representations of 3D shapes, and several dimension-specific data structures and modeling operators have been proposed in the literature. Here, we propose basic topological modeling operators for building and updating cell complexes in arbitrary dimensions. These operators either preserve the topology of the cell complex, or they modify it in a controlled way. We compare these operators with the existing ones proposed in the literature, in particular with handle operators and various Euler operators on 2D and 3D cell complexes.

Keywords: geometric modeling, cell complexes, topology-preserving operators, topology-modifying operators.

1 Introduction

Cell complexes, together with simplicial complexes, have been used as a modeling tool in a wide range of application domains, such as solid modeling, computer graphics, computer aided design, finite element analysis, animation, scientific visualization, and geographic data processing. They are used to discretize geometric shapes, such as static and dynamic 3D objects, or surfaces and hyper-surfaces describing the behavior of scalar or vector fields.

The literature on operators for building and updating cell complexes is vast but quite disorganized. A variety of topological operators have been designed for building and updating data structures representing 2D and 3D cell complexes, such as handle operators and Euler operators. Handle operators are based on handlebody theory, stating that any n-manifold can be obtained from an n-ball by attaching handles to it. The main characteristic of Euler operators is that they maintain the Euler-Poincaré formula expressing the topological validity condition of a cell complex.

We propose a set of Euler operators which form a minimally complete basis for building and updating cell complexes in arbitrary dimensions in a topologically consistent manner. We distinguish between operators that maintain the topology

M. Ferri et al. (Eds.): CTIC 2012, LNCS 7309, pp. 98–107, 2012.

of the complex, and the ones that modify it in a controlled manner. Topology-preserving operators add (or remove) a pair of cells of consecutive dimension, but they do not change the Betti numbers of the complex. Topology-modifying operators add (or remove) an i-cell, and increase (decrease) the ith Betti number, or decrease (increase) the $(i-1)$st Betti number of Γ. We compare the 2D and 3D instance of our operators with other Euler operators, and with handle operators, proposed in the literature for 2D and 3D cell complexes, and we show how these latter can be expressed in terms of the Euler operators we define here.

In Section 2, we review some background notions on cell complexes. In Section 3, we introduce a set of topological operators for building and updating cell complexes in arbitrary dimensions. In Section 4 we show how Euler operators proposed in the literature can be expressed through 2D and 3D instances of our operators, and in Section 5, we show how handle operators can be expressed in terms of our operators. In Section 6, we draw some concluding remarks, and discuss possible research directions.

2   Background Notions

We review here some notions on cell complexes, that we will use throughout this paper (see [Ago05] for more details). A k-cell in the Euclidean space $\mathbb{E}^n$ is a homeomorphic image of a k-dimensional ball, and a cell complex in $\mathbb{E}^n$ is a finite set Γ of cells in $\mathbb{E}^n$ of dimension at most d, $0 \leq d \leq n$, such that

- the cells in Γ are pairwise disjoint,
- for each cell $\gamma \in \Gamma$, the boundary of γ is a disjoint union of cells of Γ.

If the maximum dimension of the cells in Γ is equal to d, then Γ is called a d-complex. The set of cells on the boundary of a cell γ is called the (combinatorial) boundary of γ. The (combinatorial) co-boundary (or star) of γ consists of all the cells of Γ that have γ on their combinatorial boundary. An h-cell γ' on the boundary of a k-cell γ, $0 \leq h \leq k$, is called an h-face of γ, and γ is called a coface of γ'. Each cell γ is a face of itself. If $\gamma' \neq \gamma$, then γ' is called a proper face of γ, and γ and γ' are said to be incident. The domain, or carrier, of a cell d-complex Γ embedded in E^n, with $0 \leq d \leq n$, is the subset of E^n defined by the union, as point sets, of all the cells in Γ.

The Euler-Poincaré formula expresses the necessary validity condition of a cell complex with manifold or non-manifold carrier [Ago05]. The Euler-Poincaré formula for a cell d-complex Γ (with or without boundary, of homogenous or non-homogenous dimension) with n_i i-cells states that

$$\sum_{i=0}^{d}(-1)^i n_i = n_0 - n_1 + .. + (-1)^d n_d = \sum_{i=0}^{d}(-1)^i \beta_i = \beta_0 - \beta_1 + .. + (-1)^d \beta_d.$$

Here, β_i is the ith Betti number of Γ, and it measures the number of independent non-bounding i-cycles in Γ, i.e., the number of independent i-holes. The alternating sum $n_0 - n_1 + .. + (-1)^d n_d$ is denoted as $\chi(\Gamma)$, and is called the Euler-Poincaré characteristic of Γ.

3   Topological Operators

There have been many diverse proposals in the literature for manipulation operators on 2D and 3D cell complexes. We propose here a minimal set of Euler operators on cell complexes in arbitrary dimensions, which subsume all the other Euler operators proposed in the literature. We classify these operators as:

- topology-preserving operators: $MiC(i + 1)C$ (*Make i-Cell and (i+1)-Cell*),
- topology-modifying operators: $MiCiCycle$ (*Make i-Cell and i-Cycle*).

There are in total d topology-preserving operators, and $(d+1)$ topology-modifying operators.

Topology-preserving operators $MiC(i + 1)C$ change the number of cells in the complex Γ, by increasing the number n_i of i-cells and the number n_{i+1} of $(i + 1)$-cells by one. The inverse $KiC(i + 1)C$ (*Kill i-Cell and (i+1)-Cell*) operators delete an i-cell and an $(i + 1)$-cell from Γ. The Euler characteristic and the Betti numbers of the complex remain unchanged. Topology-preserving operator $MiC(i+1)C$ can create two new cells from an existing i- or $(i+1)$-cell, or insert the new cells in the complex. The first type of $MiC(i + 1)C$ operator either splits an existing $(i+1)$-cell into two by splitting its boundary, and creates an i-cell separating the two $(i + 1)$-cells, or dually, it splits an existing i-cell in two by splitting its co-boundary, and creates an $(i + 1)$-cell bounded by the two i-cells. The second type of $MiC(i+1)C$ operator either creates an i-cell and an $(i + 1)$-cell bounded only by the i-cell, or dually, it creates an $(i + 1)$-cell and an i-cell bounding only the $(i + 1)$-cell.

Topology-modifying operators change both the number of cells in the complex Γ and its Betti numbers, and thus they change the Euler characteristic of Γ. They increase the number n_i of i-cells and the number β_i of non-bounding i-cycles by one. The inverse $KiCiCycle$ (*Kill i-Cell and i-Cycle*) operators delete an i-cell and destroy an i-cycle, thus decreasing the numbers n_i and β_i by one.

We note here that the Betti numbers of a cell complex are determined by the immediate boundary relation ∂_i on the complex, which relates i-cells with $(i-1)$-cells. The creation of an i-cell affects relations ∂_{i+1} and ∂_i, and thus it either increases β_i by one, or it decreases β_{i-1} by one, to maintain the validity of the Euler-Poincaré formula. In the first case, we obtain our operators $MiCiCycle$ (Make i-Cell and i-Cycle).

Operators $MiCK(i-1)Cycle$ (Make i-Cell, Kill $(i-1)$-Cycle), $i \geq 2$, obtained in the second case, can be expressed through the proposed ones as: $K(i-1)C(i-1)Cycle$ (Kill $(i - 1)$-Cell and $(i - 1)$-Cycle) applied on one $(i - 1)$-cell on the boundary of the i-cell to be created (possibly preceded by some $K(i - 1)CiC$, which delete all i-cells in the co-boundary of the $(i - 1)$-cell), followed by $M(i-1)CiC$ (Make $(i - 1)$-Cell and i-Cell), which re-creates the deleted $(i - 1)$-cell and creates the i-cell (followed by $M(i - 1)CiC$, inverse to $K(i - 1)CiC$, which restore the deleted cells).

$M1CK0Cycle$ (Make 1-cell, Kill 0-Cycle) can be obtained by destroying one of the merged 0-cycles (components), applying one $M0C1C$, and re-creating the destroyed component.

$M0C0Cycle$ (Make 0-Cell and 0-Cycle, i.e., make vertex and connected component) is also an *initialization* operator, which creates a new complex Γ.

It can be shown that the proposed operators form a complete set of basis operators for creating and updating cell complexes, by interpreting these operators as ordered $(2d+2)$-tuples $(x_0, x_1, .., x_d, c_0, c_1, .., c_d)$ in an integer grid, belonging to the hyperplane $\sum_{i=0}^{d}(-1)^i x_i = \sum_{i=0}^{d}(-1)^i c_i$ defined by the Euler-Poincaré formula. The first $d+1$ coordinates denote the number of i-cells created or deleted by the operator, depending on the sign of the coordinate, and the last $d+1$ coordinates denote the change in the Betti numbers of the complex induced by the operator. Operator $MiC(i+1)C$, $0 \leq i \leq d-1$, has coordinates $x_i = x_{i+1} = 1$, $x_j = 0$, $j \in \{0, 1, ..., d\}\setminus\{i, i+1\}$, $c_j = 0$, $j \in \{0, 1, .., d\}$. Operator $MiCiCycle$, $0 \leq i \leq d$, has coordinates $x_i = c_i = 1$, $x_j = c_j = 0$, $j \in \{0, 1, ..., d\}\setminus\{i\}$.

A linear combination $\sum_{i=0}^{d}\alpha_d MiC(i+1)C + \sum_{i=0}^{d}\beta_d MiCiCycle$ vanishes if and only if $\alpha_i = \beta_i = 0$, $0 \leq i \leq d$, implying that the tuples corresponding to our operators are linearly independent. Each tuple $(a_0, a_1, .., a_d, b_0, b_1, .., b_d)$ in the hyperplane can be expressed through the $2d+1$ independent $(2d+2)$-tuples corresponding to our operators as $\sum_{i=0}^{d}\alpha_d MiC(i+1)C + \sum_{i=0}^{d}\beta_d MiCiCycle$, where $\alpha_i = \sum_{j=0}^{i}(-1)^{i-j}a_j - \sum_{j=0}^{i}(-1)^{i-j}b_j$, and $\beta_i = b_i$, $0 \leq i \leq d$. Thus, each operator satisfying the Euler-Poincaré formula on a cell complex Γ can be expressed through our operators. In the space (hyperplane) of dimension $(2d+1)$, a generating set consisting of $(2d+1)$ independent tuples forms a basis for the hyperplane.

For a 2-complex Γ embedded in $\mathbb{E}^3$, the operators are also called:

- Topology-preserving operators: MVE (*Make Vertex and Edge*) and MEF (*Make Edge and Face*).
- Topology-modifying operators: $MV0Cycle$ (*Make Vertex and 0-Cycle*), $ME1Cycle$ (*Make Edge and 1-Cycle*) and $MF2Cycle$ (*Make Face and 2-Cycle*).

Operator $MV0Cycle$ creates a new vertex and a new connected component, it increases by one the number of vertices (0-cells) and the zeroth Betti number β_0. It is also an initialization operator. Operator $ME1Cycle$ creates a new edge and forms a 1-cycle, thus increasing by one the number of edges (1-cells) and the first Betti number β_1. Operator $ME2Cycle$ creates a new face and forms a 2-cycle, thus increasing by one the number of faces (2-cells) and the second Betti number β_2. Figure 1 shows an example of $MV0Cycle$, $ME1Cycle$ and $MF2Cycle$ operators.

For a 3-complex Γ embedded in $\mathbb{E}^3$, there will be an additional topology-preserving operator $MFVl$ (*Make Face and Volume (3-Cell)*) which adds a new face (2-cell) and a new three-dimensional (volumetric) cell. The topology-modifying operators will be the same as for 2-complexes, since in this case the third Betti number β_3 is null.

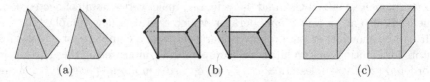

Fig. 1. Topology-modifying operators on a 2-complex in $\mathbb{E}^3$: $MV0Cycle$ (*Make Vertex and 0-Cycle*) (a); $ME1Cycle$ (*Make Edge and 1-Cycle*) (b); $MF2Cycle$ (*Make Face and 2-Cycle*) (c)

4   Comparison with Other Euler Operators

We show that various Euler operators proposed in the literature for 2D and 3D cell complexes are either instances of our operators, or can be expressed in terms of them.

Virtually all proposed sets of basis Euler operators contain MEV (Make Edge and Vertex) and MEF (Make Edge and Face) operators, which are the same as our $M0C1C$ (Make 0-cell and 1-cell) and $M1C2C$ (Make 1-cell and 2-cell) operators, respectively.

Several sets of basis operators have been proposed for manifold 2-complexes bounding a solid in $\mathbb{E}^3$, called *boundary models*. In these models, there is only one implicitly represented volumetric cell (corresponding to the cavity determined by the complex), which is not necessarily a topological cell. Various forms of topology-modifying operators [EW79, BHS80, MS82, Man88] are defined for such models.

The *glue* operator in [EW79] merges two faces and deletes both of them. It corresponds to the connected sum operator on manifold surfaces. Two faces may be glued if they have the same number of vertices, and they have no edges in common. The glue operator deletes not only the two faces, but it deletes also all the edges and vertices on the boundary of one of the deleted faces. There are two instances of the glue operator, illustrated in Figure 2.

- if the two glued faces belong to two different shells, one shell is deleted (β_0 is decreased by one), and the operator is called KFS (*Kill Face and Shell*).
- if the two glued faces belong to the same shell, a handle (genus) is created (β_1 is increased by two), and the operator is called $KFMH$ (*Kill Face, Make Hole*).

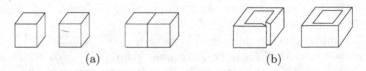

Fig. 2. Two instances of the glue operator: KFS (*Kill Face and Shell*) (a), $KFMH$ (*Kill Face, Make Hole*) (b)

Let $v_1, e_1, v_2, e_2, .., v_k, e_k$ and $v_1', e_1', v_2', e_2', .., v_k', e_k'$ be the cyclical lists of vertices and edges of the two glued faces f and f', listed in the order in which they are identified. Both instances of the glue operator can be expressed through our operators as follows:

- $M1CK0Cycle$ for KFS and $M1C1Cycle$ for $KFMH$ creates an edge connecting v_1 and v_1',
- $K0C1C$ contracts the edge (v_1, v_1'), and identifies v_1' with v_1 (vertex v_1 is the current vertex),
- $M1C2C$ makes a triangular face with vertices v_i, v_{i+1}, v_{i+1}' for current vertex v_i, $K0C1C$ and $K1C2C$ identify vertex v_{i+1}' with vertex v_{i+1} and edge e_{i+1}' with edge e_{i+1}, respectively (v_{i+1} is the current vertex),
- $M2C2Cycle$ and $K1C2C$ identify edge e_k' with edge e_k,
- $K2C2Cycle$ deletes face f' and the 2-cycle formed by faces f and f',
- $K2C2Cycle$ for KFS merges the two solids bounded by shells containing f and f' into one, and $K2CM1Cycle$ for $KFMH$ deletes face f and creates a 1-cycle determined by edges $e_1,..,e_k$.

In [BHS80, MS82, Man88], the topology-modifying operator is called $MRKF$ (*Make Ring, Kill Face*). It is similar to the glue operator in [EW79], but it imposes less restrictive conditions on the glued faces, and it deletes only one of the faces. It creates a ring and deletes a face from the model, by gluing a face to another face, thus deleting one face and making an (inner) ring in another face. The face that is not deleted is transformed into a non topological cell. The operator has two instances:

- $MRKFS$ (*Make Ring, Kill Face and Shell*) glues together two faces belonging to two different shells, thus merging two shells into one.
- $KFMRH$ (*Kill Face, Make Ring and Hole*) glues two faces belonging to the same shell, thus making a hole (genus) in the surface.

Let f' be the face that is glued to face f, and deleted. $MRKF$ can be expressed through our operators as follows (see Figure 3):

- k $M0C1C$ operators, where k is the number of edges and of vertices of f',
- k $M1C2C$ operators, which create a copy of f' in f,
- a sequence defining the glue operator in [EW79],
- $(k-1)$ $K1C2C$ operators (we leave one edge joining a vertex of f to a vertex of f' to maintain the topological validity of face f).

Topology-modifying operators defined for non-manifold 2-complexes in $\mathbb{E}^3$ [LL01] are called $MECh$ (*Make Edge and Complex Hole*), $MFKCh$ (*Make Face, Kill Complex Hole*) and $MFCc$ (*Make Face and Complex Cavity*). They are the same as our operators $M1C1Cycle$, $M2CK1Cycle$ and $M2C2Cycle$, respectively. For 3-complexes in $\mathbb{E}^3$ [MSNK89, Mas93], an additional topology-modifying operator is defined, called $MVlKCc$ (*Make Volume, Kill Complex Cavity*), which is the same as our $M3CK2Cycle$ operator.

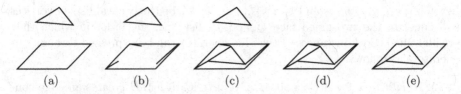

Fig. 3. $MRKF$ operator expressed through our operators: triangular face to be glued to the quadrangular face (a), three $M0C1C$ operators (b), three $M1C2C$ operators (c), *glue* (d), two $K1C2C$ operators (e)

5   Comparison with Handle Operators

Handle operators on a manifold cell 2-complex Γ triangulating a surface S have been introduced in [LPT+03]. They are based on handlebody theory for surfaces [Mil63, Mat02], stating that any surface S can be obtained from a 2-ball by iteratively attaching handles (0-, 1- and 2-handles).

Attachment of a 0-handle is also an initialization operator. It creates a new surface with one face, three edges and three vertices. There are three operators that correspond to attaching a 1-handle. They identify two boundary edges of Γ (incident in exactly one face) with no vertices in common. If the two identified edges belong to two different components of Γ, then the number of connected components and of boundary curves (connected components of boundary edges) in Γ is decreased by one. If the two identified edges belong to the same component and the same boundary curve of Γ, then the number of holes (independent 1-cycles) and the number of boundary curves in Γ is increased by 1. If the two identified edges belong to the same component and two different boundary curves of Γ, then the number of holes (independent 1-cycles) is increased by 1, and number of boundary curves in Γ is decreased by 1. The operator that corresponds to the attachment of a 2-handle identifies two edges on the boundary of Γ with two vertices in common. It decreases the number of holes and the number of boundary curves in Γ by 1. Handle operators in 2D are illustrated in Figure 4.

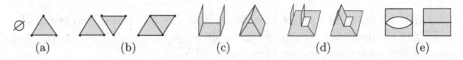

Fig. 4. Handle operators in 2D: attachment of a 0-handle (a); attachment of a 1-hande (b), (c) and (d); attachment of a 2-handle (e)

Handle operators can be classified as topology-modifying operators, and they can be expressed through our operators as discussed below:

– The attachment of a 0-handle corresponds to creating an initial triangle (a 2-ball). It can be expressed as $M0C0Cycle$ operator, two $M0C1C$ operators and one $M1C2C$ operator, which together create a triangle (see Figure 5 (a)).

– The attachment of a 1-handle identifies two boundary edges e_1 and e_2 with no vertices in common. It can be expressed through one $M1CK0Cycle$ and one $M1C1Cycle$ operator if e_1 and e_2 belong to different components, or two $M1C1Cycle$ operators if they belong to the same component (the created edges connect the endpoints of e_1 to the corresponding endpoints of e_2), two $K0C1C$ operators (they contract the two created edges and identify the corresponding endpoints), one $M2CK1Cycle$ operator (it creates a face that fills the ring and deletes the cycle formed by e_1 and e_2), and finally one $K1C2C$ operator (it contracts the created face and identifies e_1 with e_2) (see Figure 5 (b)).

– The attachment of a 2-handle identifies two edges with both vertices in common. It can be expressed as a $M2CK1Cycle$ operator, followed by $K1C2C$ operator.

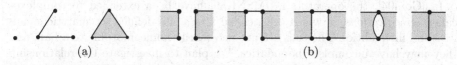

(a)   (b)

Fig. 5. Attachment of a 0-handle in 2D can be expressed as one $M0C0Cycle$, two $M0C1C$ and one $M1C2C$ (a). Attachment of a 1-handle in 2D can be expressed as one $M1CK0Cycle$ or one $M1C1Cycle$, one $M1C1Cycle$, two $K0C1C$, one $M2CK1Cycle$ and one $K1C2C$ (b).

Handle operators have been extended to 3D in [LT97]. The operator that creates a new 3-ball (initialization operator) corresponds to the attachment of a 0-handle. Other operators identify two boundary faces (incident in exactly one 3-cell) of a cell 3-complex Γ triangulating a solid S. The attachment of a 1-handle can be applied in three situations: if the two identified boundary faces are on different connected components of Γ, then the two components are merged into one; if the two identified faces belong to the same boundary surface component of Γ (connected component of boundary faces) and have no edges in common, then a hole is created; if the two identified faces belong to the different boundary surfaces of the same connected component of Γ, the operator can be realized only if Γ is embedded in a space of dimension greater than 3. The attachment of a 2-handle corresponds to identifying two faces on the same boundary surface component of Γ that have some edges in common. The operator can create cavities and/or close holes in Γ. The attachment of a 3-handle is applicable if the two identified faces belong to the same boundary surface component and have all edges in common. This operator fills in the cavity formed by the two identified faces.

The handle operators in 3D generalize the glue operator in [EW79], since the two faces identified by a handle operator may have none, some, or all edges in common. Thus, they can be expressed in terms of our operators in a similar manner.

6 Concluding Remarks

We have introduced a complete set of basis operators for building and updating cell complexes in arbitrary dimensions. We have shown how other Euler operators and handle operators proposed in the literature on 2D and 3D cell complexes can be expressed through our operators.

An interesting research direction would be to generalize handle operators to higher dimensions, i.e., for an n-manifold discretized as a cell complex, and to express them as macro-operators through Euler operators in higher dimensions.

Simplicial complexes, which can be considered as a special case of cell complexes, have been used as a preferred representation of shapes long before cell complexes. In [LLM+10], a unified framework for building and updating manifold cell 2-complexes has been proposed. It combines topology-preserving stellar operators, and topology-modifying handle operators. We believe that it would be interesting to combine our topology-modifying operators with stellar operators.

In [Gom04], the operators in [MSNK89] have been extended to complexes called *stratifications*, in which cells, called *strata*, are defined by analytic equalities and inequalities. The cells are not necessarily homeomorphic to a ball, and they may have incomplete boundaries. We plan to investigate the relationship between our operators and the ones in [Gom04], in the case of cell complexes.

Hierarchical pyramidal models have been defined in the framework of combinatorial maps [BK03], based on topology-preserving operators. We plan to define a multi-resolution model for cell complexes based on our operators.

Acknowledgments. This work has been partially supported by the Italian Ministry of Education and Research under the PRIN 2009 program, and by the National Science Foundation under grant number IIS-1116747.

References

[Ago05]   Agoston, M.K.: Computer Graphics and Geometric Modeling: Mathematics. Springer-Verlag London Ltd. (2005) ISBN:1-85233-817-2

[BHS80]   Braid, I.C., Hillyard, R.C., Stroud, I.A.: Stepwise construction of polyhedra in geometric modelling. In: Mathematical Methods in Computer Graphics and Design, pp. 123–141. Academic Press (1980)

[BK03]   Brun, L., Kropatsch, W.G.: Receptive fields within the Combinatorial Pyramid framework. Graphical Models 65(1-3), 23–42 (2003)

[EW79]   Eastman, C.M., Weiler, K.: Geometric Modeling Using the Euler Operators. In: 1st Annual Conference on Computer Graphics in CAD/CAM Systems. MIT (May 1979)

[Gom04]   Gomes, A.J.P.: Euler operators for stratified objects with incomplete boundaries. In: Proceedings of the Ninth ACM Symposium on Solid Modeling and Applications, SM 2004, pp. 315–320. Eurographics Association (2004)

[LL01]   Lee, S.H., Lee, K.: Partial entity structure: a fast and compact non-manifold boundary representation based on partial topological entities. In: Proceedings Sixth ACM Symposium on Solid Modeling and Applications, pp. 159–170. Ann Arbor, Michigan (2001)

[LLM+10]   Lewiner, T., Lopes, H., Medeiros, E., Tavares, G., Velho, L.: Topological
           mesh operators. Computer Aided Geometric Design 27(1), 1–22 (2010)

[LPT+03]   Lopes, H., Pesco, S., Tavares, G., Maia, M., Xavier, A.: Handlebody Rep-
           resentation for Surfaces and Its Applications to Terrain Modeling. Inter-
           national Journal of Shape Modeling 9(1), 61–77 (2003)

[LT97]   Lopes, H., Tavares, G.: Structural Operators for Modeling 3-Manifolds. In:
           Proceedings Fourth ACM Symposium on Solid Modeling and Applications,
           pp. 10–18. ACM Press (May 1997)

[Man88]   Mantyla, M.: An Introduction to Solid Modeling. Computer Science Press
           (1988)

[Mas93]   Masuda, H.: Topological Operators and Boolean Operations for Complex-
           Based Non-Manifold Geometric Models. Computer-Aided Design 25(2),
           119–129 (1993)

[Mat02]   Matsumoto, Y.: An Introduction to Morse Theory. Translations of Math-
           ematical Monographs, vol. 208. American Mathematical Society (2002)

[Mil63]   Milnor, J.: Morse Theory. Princeton University Press, New Jersey (1963)

[MS82]   Mantyla, M., Sulonen, R.: Gwb: A Solid Modeler with Euler Operators.
           IEEE Computer Graphics and Applications 2, 17–31 (1982)

[MSNK89]   Masuda, H., Shimada, K., Numao, M., Kawabe, S.: A Mathematical The-
           ory and Applications of Non-Manifold Geometric Modeling. In: IFIP WG
           5.2/GI International Symposium on Advanced Geometric Modeling for
           Engineering Applications, pp. 89–103. North-Holland (1989)

Triangle Mesh Compression
and Homological Spanning Forests

Javier Carnero, Helena Molina-Abril, and Pedro Real

Computational Topology and Applied Mathematics Group,
Applied Mathematics I Department, University of Seville
javier@carnero.net, {habril,real}@us.es

Abstract. Triangle three-dimensional meshes have been widely used to represent 3D objects in several applications. These meshes are usually surfaces that require a huge amount of resources when they are stored, processed or transmitted. Therefore, many algorithms proposing an efficient compression of these meshes have been developed since the early 1990s. In this paper we propose a lossless method that compresses the connectivity of the mesh by using a valence-driven approach. Our algorithm introduces an improvement over the currently available valence-driven methods, being able to deal with triangular surfaces of arbitrary topology and encoding, at the same time, the topological information of the mesh by using Homological Spanning Forests. We plan to develop in the future (geo-topological) image analysis and processing algorithms, that directly work with the compressed data.

Keywords: Triangle Mesh Compression, Homological Spanning Forest, Computational algebraic topology.

1   Introduction

Polygon three-dimensional meshes have been widely used on many different applications to represent 3D objects. In fact, since triangles are the basic geometric primitives for standard graphics hardware and for many simulation algorithms, triangle meshes are the most commonly used. This is the reason why most of the effort in the field of static 3D model compression has been devoted to triangle meshes.

We focus here in triangle meshes that often require a huge amount of data for their storage, processing and transmission in the raw data format. Therefore, to find an efficient method for compressing these meshes is one of the aims of the work we present here.

Besides that, the ability of computing and storing topological information of these meshes is also an issue of interest for many problems dealing with them. This is the case of medical image processing, where topological information is crucial in order, for instance, to make a meaningful automatic classification of the images.

M. Ferri et al. (Eds.): CTIC 2012, LNCS 7309, pp. 108–116, 2012.

The method we propose here provides a lossless compression of the connectivity of the mesh, allowing the inclusion of its topological information in the coded file and/or its automatic computation in the decompression process.

In order to compress the connectivity of the mesh, a valence-driven approach started by Touma and Gotsman [2] is used. This valence-driven approach codifies the neighborhood of each vertex as a number following a certain order. Using this "valence"information the mesh can be reconstructed without loosing information. Our variation introduces an improvement over the currently available valence-driven algorithms, being able to deal with triangular orientable meshes of arbitrary topology (not necessary 2-manifold), keeping all the benefits of the valence-driven approach.

In the compression method we propose here, the data structures defined by Molina-Abril and Real [4] (called *Homological Spanning Forest*) can be easily computed without increasing the computational time of the algorithm. Let us notice that by using this structure, we encode not only basic topological information like Betti numbers, genus or Euler characteristic, but also advanced topological information (reconstruction of the boundary, homological classification of cycles, etc.). The inclusion of this structure in the compressed data provides a suitable framework for a geo-topological processing in the compressed domain (contractibility testing and transformability of cycles, topological analysis of ROIs, shortest path problems, etc.). The possibility of directly working with the compressed data, is an important advantage when dealing with large meshes and images.

The resulting algorithm uses less than 1.5 bits per vertex (bpv) on average to encode mesh connectivity. This compression ratio coincides with the state-of-the-art ratio that has not been seriously challenged till now. As other valence-driven algorithms, the proposed method can be used in progressive transmission, which means that the mesh can be decompressed, processed and rendered during the transmission process.

In particular the investigation about connectivity compression has been developed under the project *VirSSPA'10*, in the *Hospital Universitario Virgen del Rocío*, Seville (Spain), and financed by the *Consejería de Salud de la Junta de Andalucía* and *FEDER* founds.

In order to complete the goals of the project, an application developed in C++ implementing the introduced technique has been developed. A database of more than six hundred medical images of real patients from the *Hospital Universitario Virgen del Rocío* has been used to corroborate the theoretical results.

2   Connectivity Compression Method

For compressing the connectivity of an orientable triangle mesh, a valence-driven algorithm has been developed. This approach was presented in 1998 by Touma and Gotsman [2], compressing the connectivity in terms of the neighborhood of each vertex. In the algorithm we propose here, a random triangle of the mesh is selected to be the origin of three implosions (one for each of its three vertices).

Also the triangle face is defined as *the imploded region*, being its border defined as the three edges plus the three vertices.

The implosion of vertex v is defined as the process in which v and its neighborhood faces are destroyed by collapsing (or being squeezed in) on v. An implosion is always produced by a vertex in the border of the imploded region, growing it. The number of new vertices that have been added to the implosion region border due to the implosion is registered in order to be able to reconstruct the implosion in the decompression process.

Fig. 1. An implosion (colored in yellow) from a vertex that belongs to a hole in the mesh (in green), and the growing of the imploded region (in black) surrounding the hole

This process also produces sorted list of vertices as they pass through the regular region of the mesh to the imploded region border. This distribution, based on the mesh topology, is very useful for the next step of a triangle mesh compression, that is the geometry compression (it will not be treated in this paper).

Two special cases of implosions can be distinguished when dealing with 3D triangle surfaces. Using the terminology coined by Touma and Gotsman, we call these two cases *split* and *merge*. Although these special cases are not exactly the same as those defined by Touma and Gotsman (as our method is developed from an implosion point of view), they both are produced by the same topological principles, so the notation has still sense in our case.

- SPLIT
  A *split* is produced when an implosion touches its own region border. This produces a split of the border into two, so if the imploded region has only one border, after the split it will have two.
- MERGE
  A *merge* occurs when an implosion touches another border that is different from the border where the vertex that produced the implosions belongs. In this case, a merge of the two borders needs to be done. If the imploded region has two borders, after the merge it will have only one.

3   Homological Spanning Forest Representation

Roughly speaking, topology in a discrete context helps to understand the degree of connectivity of subdivided geometric structures. For subdivided objects, homology is topology measured in terms of linear combinations (called chains) of unit elements or bricks (also called cells), and in terms of "boundary relations" describes the connectivity dependencies among these bricks. Homology depends on the ring of coefficient, and gives an algebraic answer in terms of formal sums of bricks that have no boundary (for example, closed subdivided curves or surfaces). These sums are called cycles and homology determines a representative cycle for each n-dimensional hole or homology generator the object has (connected components, tunnels, cavities, etc). In this way, homology can be considered as a specification of the contribution of each brick to the creation of the homology representative cycles.

In order to codify these connectivity information in an efficient way, we use here a graph representation called Homological Spanning Forest (or HSF for short). These hierarchical tree-like structure gives a positive and efficient answer to the problem of codifying and computing classical algebraic topological information (Euler characteristic, Betti numbers, classification and relations between cycles, etc.). A detailed explanation about the topological information that the HSF codifies, and its formal definition can be found in [5]. Relations between the HSF and Morse Theory, in [4].

We will not go here into details about the Homological Spanning Forest representation. An elementary example of a subdivided object is shown in Figure 2 for the understanding of this idea. Given a geometric graph G, its homology information can be directly captured by means of a spanning tree T of G. In fact, we transform T into a directed tree T^d by adding arrows to every edge in T, in such a way that at most one arrow comes out from each vertex. Therefore, there will be only one vertex s of G, called sink, from which no arrow comes out. In Figure 2 we interpret an arrow (e, f) in T^d from the vertex f to the vertex e as an elementary "deformation" operation, "contracting" in a continuous way the vertex f onto e through the edge (e, f) inside the object. The result of applying (no matter the order we choose) the set of homology-preserving operations represented by a red arrow in Figure 2, is a reduced structure consisting of only three bricks: the vertex e, and two loops or "edges" starting and ending at the same common vertex e (in fact, they represent the cycles $\{(c, e), (d, e), (c, d)\}$ and $\{(c, e), (c, f), (e, f)\}$ coming from (c, e) and (c, f), respectively).

The directed spanning tree T^d can be interpreted in dynamical terms, as the way in which the set of vertices of the graph is "collapsed" to a representative vertex of the connected component (in this case, the vertex e in black). These three representative cycles of homology generators (in this case, no matter of the ground ring we use but heavily dependent on the spanning tree T) are determined by the following bricks of G (called *critical*): the edges (c, f), (c, e) that belong to $G \setminus T$, and the sink vertex e that belongs T. Their integer homology groups are one copy of $\mathbb{Z}$ in dimension 0 and two copies in dimension 1. A Homological Spanning Forest representation $\mathcal{F}(G)$ for the subdivided geometric structure G is the set of trees $\mathcal{F}(G) = \{T^d, T_1, T_2\}$, where T_1 and T_2 are trees composed by only one "vertex": the original edges (c, e) and (c, f) respectively.

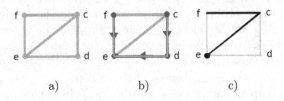

a)   b)   c)

Fig. 2. a) A geometric graph G drawn on $\mathbb{R}^2$, b) a directed spanning tree (in red) showing a homological "deformation" process, and c) the minimal homological object (in black)

If we now add two triangles to the previous graph obtaining a new object O, we can now construct a tree T^{d_2} codifying the "collapsing" of (c, f) and (c, e) to the triangles (c, e, f) and (c, d, e) respectively (see Figure 3). Then, the HSF representation $\mathcal{F}(O)$ for the subdivided object O is the set of trees $\mathcal{F}(O) = \{T^d, T^{d_2}\}$.

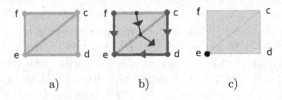

a)   b)   c)

Fig. 3. a) A geometric object O, b) a directed spanning tree (in red) showing a homological "deformation" process, and c) the minimal homological object (in black)

This graph-based representation, suitably encodes advanced topological features of the object, due to the fact that the HSF forest can be automatically rewritten in algebraic terms (with coefficients in a field) as a chain homotopy operator, that determines a strong relationship at chain level (formal sums of bricks) between the geometric object and its minimal homological expression; that is a chain homotopy equivalence. The chain homotopy operator can be directly extracted from the forest as sums of cells following the paths of the trees (see [5] for more details). By advanced topological information we mean not only Euler characteristic and Betti numbers, but also classification of cycles, relations between cycles, etc.

4   Connectivity Compression Algorithm

The compression algorithm is presented in algorithm 1. The decompression method works similarly, but reading the implosions information from the compressed file and then creating the mesh.

Algorithm 1. Compression

Require: List of triangles and vertex information (coordinates, normals and color) of an orientable triangle mesh.

Ensure: Triangle mesh compressed.

Ensure: HSF of the mesh.

 1: **while** there are no imploded vertices **do**
 2:     Select a no imploded triangle.
 3:     Define the triangle as imploded region.
 4:     Build the HSF for that triangle.
 5:     Define the triangle border as the imploded region border, and make it the active border.
 6:     **while** there is an active border **do**
 7:         **if** all vertices of the border have been imploded **then**
 8:             Delete the border.
 9:             **if** there is no inactive border **then**
10:                 End loop.
11:             **else**
12:                 Select an inactive border as active.
13:             **end if**
14:         **else**
15:             Select a not imploded vertex of the active border.
16:             Implode the vertex.
17:             Register valence, splits and merges.
18:             Build the HSF on the implosion, connecting them with the imploded region.
19:         **end if**
20:     **end while**
21: **end while**
22: **return** Homological Spanning Forest
23: **return** Compressed mesh

The Touma and Gotsman algorithm [2] has an average compression ratio of 1.5 bits per vertex. This ratio has not been seriously challenged till now. However, these results are purely empirical, and a theoretical study is not available.

Alliez and Desbrun [1] proposed a method to further improve the performance of Touma and Gotsman algorithm. They observed that the code produced by splits consumes a non-trivial portion of coding bits, and proposed some simple techniques to reduce it, especially for irregular meshes where this special code can be huge. Alliez and Desbrun proved that if the amount of the splits is negligible, the performance of their algorithm is upper-bounded by 3.24 bpv, which is exactly the same as the theoretical bits per vertex value computed by enumerating all possible planar graphs [6].

In the algorithm we present here, our results are close to the ones by Touma and Gostman. In the near future we plan to adapt the ideas presented by Alliez and Desbrun to our algorithm in order to reduce the compression ratio. This will give us an algorithm that is able to reach not only reach the theoretical minimal compression ratio, but also to include in the compressed file (by using the implosion technique) the topological information of the mesh (by computing the so called HSF data structures).

4.1   HSF Data Structures Generation

In [3], the HSF computation algorithm has quadratic time complexity. The advantage of including the HSF representation computation in the proposed compression method, is that the computational time is not increased. The forest can be directly computed at the same time the implosions are generated. On the same way, in case the compressed data do not contain the encoded HSF structure, they can also be computed during the decompression process or even directly from the compress domain.

The computation of the HSF is performed following a "star-shape" strategy (see Fig. 4). In this way when a vertex "v" implodes, the arrows going from its neighbors vertices to the vertex "v" are added to the HSF structure (colored in red in Figure 4). The first imploding vertex of the whole process is the sink vertex "s". The arrows corresponding to the neighbor triangles are also added considering the star flow (colored in blue in Figure 4). Following this strategy, the trees.

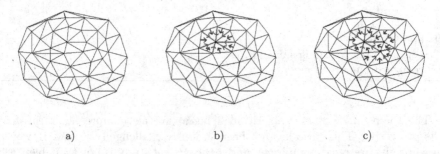

a)   b)   c)

Fig. 4. a) The initial triangle mesh, b) The first implosion and HSF generation, and c) Next steps of the compression method and HSF generation

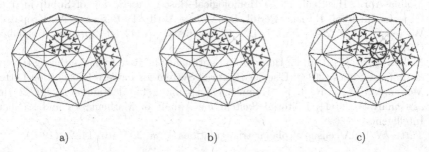

a)   b)   c)

Fig. 5. a) Two imploded areas needing to be merged, b) The first step of the HSF representation computation in the merge operation, and c) The next steps of the HSF representation computation in the merge operation producing a bifurcation in the tree (marked within a circle in the picture)

Problems may occur when a merge or split takes place (see Figure 5). In this case, the HSF structure is constructed by making a bifurcation in the T^{d_2} tree.

5   Conclusions

In this paper we propose a lossless compression method based on a valence-driven approach. The main advantage of the method, is that it compresses triangle meshes of arbitrary topology and encodes, at the same time, the topological information of the mesh by using Homological Spanning Forests, without increasing the computational time.

The topological information encoded in the HSF structure can be later used for processing geometrical and topological information in the compressed domain (automatic cycle classification, how to transform -if it is possible- a cycle into another inside the object, recognition of 3D objects based on geometrical and topological features, etc). The algorithm allows progressive transmission, in the sense that the mesh can be decompressed, processed and rendered while its data is being received.

Acknowledgements. JC, HMA and PR acknowledge the support of the project VirSSPA'10 from the *Hospital Universitario Virgen Del Rocío* founded by *Consejería de Salud de la Junta de Andalucía* and FEDER Founds, the *Computational Topology and Applied Mathematics* CATAM research group FQM-296, and the Spanish MICINN Research Project MTM2009-12716.

References

1. Alliez, P., Desbrun, M.: Valence-driven connectivity encoding for 3d meshes (2001)
2. Chen, L., Georganas, N.D.: 3d mesh compression using an efficient neighborhood-based segmentation. In: Proceedings of the 9th IEEE International Symposium on Distributed Simulation and Real-Time Applications, DS-RT 2005, pp. 78–85. IEEE Computer Society, Washington, DC (2005)

3. Molina-Abril, H., Real, P.: A Homological–Based Description of Subdivided nD Objects. In: Real, P., Diaz-Pernil, D., Molina-Abril, H., Berciano, A., Kropatsch, W. (eds.) CAIP 2011, Part I. LNCS, vol. 6854, pp. 42–50. Springer, Heidelberg (2011)
4. Molina-Abril, H., Real, P.: Homological optimality in discrete morse theory through chain homotopies (2011); Elsevier Editorial System for Pattern Recognition Letters
5. Molina-Abril, H., Real, P.: Homological spanning forest framework for 2D image analysis (2011); Editorial System for Annals of Mathematics and Artificial Intelligence
6. Tutte, W.T.: A census of planar triangulations. Can. J. Math. 14, 21 (1962)

Homology Computations via Acyclic Subspace

Piotr Brendel[1], Paweł Dłotko[1,*], Marian Mrozek[1], and Natalia Żelazna[2]

[1] Institute of Computer Science, Jagiellonian University
{piotr.brendel,pawel.dlotko,marian.mrozek}@ii.uj.edu.pl
[2] Motorola Solutions
natalia.zelazna@motorolasolutions.com

Abstract. Homology computations recently gain vivid attention in science. New methods, enabling fast and memory efficient computations are needed in order to process large simplicial complexes. In this paper we present the acyclic subspace reduction algorithm adapted to simplicial complexes. It provides fast and memory efficient preprocessing of the given data. A variant of the method for distributed computations is also presented. As a result, Betti numbers can be effectively computed.

Keywords: Homology algorithms, reduction algorithms, acyclic subspace method.

1 Introduction

The classical way of computing homology consists in finding the Smith Normal Form of the matrix of the boundary map [9]. The complexity of the Smith Normal Form algorithm is supercubical. This is prohibitive in applications where the size of the boundary map matrix is large, in particular in rigorous numerics of dynamical systems, problems in image recognition, data analysis, material science, electromagnetism, robotics, theoretical computer science, ecology, molecular biology and other areas. Therefore, in recent years several methods have been proposed to speed up homology computations, particularly computations of the homology of sets in various representations. Among such methods are geometric reduction algorithms. They aim at finding a smaller representation with the same homology as the original set. A method of this type, recently proposed in [8], is based on the construction of an acyclic subset. We recall that a simplicial complex $\mathcal{A}$ is acyclic iff the homology of $\mathcal{A}$ is isomorphic to the homology of a point. In this paper a *component-wise acyclic* subcomplex $\mathcal{A}$ of a simplicial complex $\mathcal{S}$ is a simplicial complex whose connected components are acyclic subsets of corresponding connected components of $\mathcal{S}$. The acyclic subspace algorithm is based on the simple observation that if $\mathcal{A}$ is an acyclic subcomplex of a simplicial complex $\mathcal{S}$, then:

$$H_n(\mathcal{S}) \cong \begin{cases} H_n(\mathcal{S}, \mathcal{A}) & \text{for } n \geq 1 \\ \mathbb{Z} & \text{for } n = 0 \end{cases}$$

---

* Corresponding author.

M. Ferri et al. (Eds.): CTIC 2012, LNCS 7309, pp. 117–127, 2012.

Note that due to the excision theorem [9] the relative homology $H_n(\mathcal{S}, \mathcal{A})$ depends only on the neighborhood of $\mathcal{S} \setminus \mathcal{A}$ in $\mathcal{S}$. Therefore, if a large acyclic subcomplex $\mathcal{A}$ of $\mathcal{S}$ may be constructed quickly, the problem of finding the homology of $\mathcal{S}$ is reduced to a relatively small set and consequently may be found quickly. The aim of [8] was to show that in the case of cubical complexes a relatively large acyclic subcomplex may be found in linear time.

The goal of the presented paper is to extend the ideas of [8] to the case of simplicial complexes. In particular, we propose two fast algorithms constructing a possibly large acyclic subcomplex $\mathcal{A}$ of every connected component of a simplicial complex $\mathcal{S}$. Moreover, we show how to extend this algorithms for the purposes of distributed computations.

Acyclic subset reduction leads to more efficient computation of Betti numbers, useful in image recognition.

2   Preliminaries

For the purposes of this paper a finite family of finite sets $\mathcal{S}$ is called an *abstract simplicial complex* if for every $P \in \mathcal{S}$ and for every $Q \subset P$ we have $Q \in \mathcal{S}$. An element $P \in \mathcal{S}$ is called a *simplex*. If $P \in \mathcal{S}$ and $Q \subset P$ then Q is called a *face* of P. A simplex $P \in \mathcal{S}$ is said to be *maximal* if there is no simplex $Q \in \mathcal{S}$ such that $P \subsetneq Q$. Throughout this paper $S_{max}(\mathcal{S})$ denotes the set of maximal simplices of $\mathcal{S}$. The *algebraic closure* of a simplex P, denoted by $cl(P)$ is a family of simplices consisting of P and all its faces. The closure of a family of simplices $\mathcal{K}$ is $cl(\mathcal{K}) = \bigcup_{P \in \mathcal{K}} cl(P)$. For a simplex $Q \in \mathcal{S}$ its *neighborhood* consists of all maximal simplices in $\mathcal{S}$ whose intersection with Q is nonempty . We denote this set by

$$n(Q) = \{ P \in \mathcal{S} \mid Q \cap P \neq \emptyset \text{ and } P \text{ is a maximal in } \mathcal{S} \}.$$

By the dimension of a simplex P we mean $dim(P) := card(P) - 1$. For a simplicial complex $\mathcal{S}$ by $S_0(\mathcal{S})$ we denote the set of all the vertices of $\mathcal{S}$, i.e. its 0-dimensional simplices, and we make a technical assumption that every vertex in S_0 has a unique label. In the sequel, we use a hash table [1], denoted H, whose keys are labels of vertices and for each key the value is the list of all maximal simplices containing the vertex labeled with the given key.

The main homological tools used in the paper are the exact sequence of a pair and the Mayer-Vietoris sequence [9]. The exact sequence of a pair is used to conclude that a subcomplex with trivial reduced homology can be removed from the initial complex without changing its reduced homology. From the Mayer-Vietoris sequence it follows, that a simplex can be added to the constructed acyclic subcomplex if and only if its intersection with the acyclic subcomplex has trivial reduced homology.

3   Incidence Graph

We say that a graph $G = (V, E)$ is an *incidence graph* of a simplicial complex $\mathcal{S}$ if V is the set of maximal simplices of $\mathcal{S}$ and $(S_1, S_2) \in E$ iff $S_1 \cap S_2 \neq \emptyset$.

An *augmented incidence graph* is a triple (V, E, C) where (V, E) is the incidence graph and C is the list of connected components of incidence graph, in which each connected component is represented by a single maximal simplex from this component. We will use augmented incidence graphs to retrieve all the information about neighborhoods in a simplicial complex, necessary in the process of constructing an acyclic subset.

In this section we show an algorithm constructing such a graph for a given simplicial complex. The input data for this algorithm is the list of maximal simplices $S_{max}(\mathcal{S})$ and VertexHash H, described in Section 2. For each vertex v we consider the list $H[v]$ storing the maximal simplices that contain v. $\mathcal{Q}$ denotes the queue used to store simplices which have not yet been added to the incidence graph and whose neighbors are already there. Functions **Enqueue** and **Dequeue** are standard operations on queues and their description can be found in [1].

Algorithm 3.1. IncidenceGraph(MaximalSimplexList $S_{max}(\mathcal{S})$, VertexHash H)

1: $V := \emptyset$; $E := \emptyset$; $C := \emptyset$; $\mathcal{Q} :=$ EmptyQueue;
2: **for all** Simplex $P \in S_{max}(\mathcal{S})$ **do**
3:     **if** $P \notin V$ **then**
4:         $C := C \cup \{P\}$;
5:         Enqueue($\mathcal{Q}, P$);
6:         **while** $\mathcal{Q} \neq \emptyset$ **do**
7:             Simplex *current* := Dequeue($\mathcal{Q}$);
8:             $V := V \cup \{current\}$;
9:             **for all** Vertex $v \in current$ **do**
10:                 **for all** Simplex *neighbour* $\in H[v]$, *neighbour* $\neq$ *current* **do**
11:                     **if** *neighbour* $\notin V$ **then**
12:                         $e := (current, neighbour)$; $E := E \cup \{e\}$;
13:                         **if** *neighbour* $\notin \mathcal{Q}$ **then**
14:                             Enqueue($\mathcal{Q}, neighbour$);
15: return Graph(V, E, C);

Theorem 3.1. Algorithm 3.1 stops and constructs the augmented incidence graph $G = (V, E, C)$ for simplicial complex $\mathcal{S}$ in $O(card(V) \cdot dim(\mathcal{S}) \cdot deg(H))$ where $dim(S) = max_{P \in \mathcal{S}}\{dim(P)\}$ and $deg(H) = max_{v \in S_0(\mathcal{S})}\{length(H[v])\}$. Moreover, for each connected component $G' \subset G$ its set of nodes $V(G')$ equals to the set of maximal simplices in the corresponding connected component $\mathcal{S}' \subset \mathcal{S}$.

Proof. Obviously V contains all maximal simplices in $S_{max}(\mathcal{S})$. Pair $(S_1, S_2) \in E$ iff $S_1 \cap S_2 \neq \emptyset$, therefore augmented incidence graph is obtained. Simplex P is added to C in line 4 only if $P \notin V$ which means $P \cap S = \emptyset$ for all $S \in V$ and P represents a new connected component, since in **while** loop at line 6 BFS procedure, which finds connected components, is implemented. Simplex P is added to $\mathcal{Q}$ only once, hence the **while** loop in line 6 always completes after adding to V all elements from connected component of P.

The internal **for all** loop in line 9 is performed for every d-dimensional simplex at most $dim(\mathcal{S}) \cdot deg(H)$ times. Since the **while** loop in line 6 is performed at most $card(V)$ times, the complexity of the algorithm is $O(card(V) \cdot dim(\mathcal{S}) \cdot deg(H))$. □

4 Constructing Acyclic Subset

Since the homology of a disconnected complex is a direct sum of homologies of its connected components, later in this paper we will construct component-wise acyclic subcomplexes. In this section we present two approaches to construction of such complexes using the incidence graph as well as a function named AcyclicityTest. The purpose of this function is to decide whether a simplex may be added to the constructed acyclic set without loosing acyclicity. How to obtain such a function is the purpose of Section 6.

The first algorithm, referenced in the following sections as AccST, is an adaptation of the algorithm presented in [8] to the case of simplicial complexes. The adaptation is not straightforward, because, unlike the case of cubical sets, in the simplicial case it is not obvious how to efficiently determine the neighborhood of a simplex. For this, we use the incidence graph presented in Section 3. Another difference is that, instead of one, we construct several acyclic subsets in each connected component of $\mathcal{S}$ and then join them by a spanning tree. It allows us to construct larger acyclic subsets for certain kinds of data. To do this we need two auxiliary functions: FindSimplexNotInAccSub and CreateSpanningTree. Both are based on standard graph algorithms [1]. The first finds a simplex that has no intersection with the acyclic subset. It uses breadth-first search algorithm. If no such simplex can be found, it returns NULL. The other takes as an input a list of simplices, one per each constructed acyclic subset. It first constructs the shortest paths connecting the acyclic subsets. Each path is a list of one-dimensional simplices. The paths are used to build a graph in which nodes are the constructed disjoint acyclic subsets and edges are the constructed paths. Then, Kruskal algorithm [1] is applied to create a spanning tree joining the constructed acyclic subsets.

Theorem 4.1. Algorithm 4.1 always stops. Given a simplicial complex $\mathcal{S}$ on input, represented by the incidence graph $G = (V, E, C)$, it returns a component-wise acyclic complex $\mathcal{A}$ on output.

Proof. In lines 11 and 12 a simplex P is added simultaneously to Q and to the acyclic subset $\mathcal{A}$. Since P may be added to $\mathcal{A}$ only once and the number of simplices is finite, the inner **while** loop in line 7 always finishes. Functions FindSimplexNotInAccSub and CreateSpanningTree are respectively BFS and Kruskal algorithms [1], hence they both complete. Every simplex found by FindSimplexNotInAccSub in line 13 is added to the acyclic subset. Hence, the finiteness of V implies that the **while** loop in line 4 completes. Thus, since the number of simplices in C is also finite, we know that the algorithm always stops.

Algorithm 4.1. AccST(IncidenceGraph (V, E, C))

1: $\mathcal{A} := \emptyset$; $\mathcal{Q} :=$ EmptyQueue;
2: **for all** Simplex $P \in C$ **do**
3:     $\mathcal{L} :=$ EmptyList;
4:     **while** $P \neq \emptyset$ **do**
5:         $\mathcal{A} := \mathcal{A} \cup \{P\}$;
6:         Enqueue($\mathcal{Q}, P$);
7:         **while** $\mathcal{Q} \neq \emptyset$ **do**
8:             Simplex $Q :=$ Dequeue($\mathcal{Q}$);
9:             **for all** Simplex $S \in n(Q) \setminus \mathcal{A}$ **do**
10:                 **if** AcyclicityTest($\mathcal{A}, S$) = true **then**
11:                     $\mathcal{A} := \mathcal{A} \cup \{S\}$;
12:                     Enqueue($\mathcal{Q}, S$);
13:         $P :=$ FindSimplexNotInAccSub($V, E, P, \mathcal{A}$);
14:         **if** $P \neq$ NULL **then**
15:             $\mathcal{L} := \mathcal{L} \cup \{P\}$;
16:     $\mathcal{A} := \mathcal{A} \cup$ CreateSpanningTree($\mathcal{L}$);
17: **return** $\mathcal{A}$;

Since each simplex is acyclic, we begin the construction of the acyclic subsets of the components of $\mathcal{S}$ with the representants of the connected components of the incidence graph described in Section 3. As long as we can find a simplex acyclically intersecting $\mathcal{A}$, by Mayer-Vietoris Theorem we may add it to $\mathcal{A}$ without losing its acyclicity. If we cannot find such a simplex, we look in the same connected component for another one that has no intersection with $\mathcal{A}$ and we build acyclic subset around it as described above. We stop this procedure when there are no simplices that do not intersect $\mathcal{A}$. Due to Mayer-Vietoris Theorem acyclic subsets constructed that way cannot intersect each other. Now let us assume, we have a number of disjoint acyclic subsets of $\mathcal{A}$ and we want to connect them in order to form a larger acyclic subset. First, we need to find paths, i.e. lists of one-dimensional simplices, joining the subsets. Since all components of the constructed set $\mathcal{A}$ are contained in the same connected component of $\mathcal{S}$, we can always find a path joining any two of them. Unfortunately, the constructed paths can intersect each other or even other parts of $\mathcal{A}$ creating unwanted cycles. Nevertheless, is is not difficult to avoid this problem by joining acyclic parts step by step and adding only parts of the connecting paths so as not to lose acyclicity. $\square$

Algorithm 4.2, referenced in the following sections as AccIG, constructs simultaneously the incidence graph and a component-wise acyclic complex. For certain kinds of data it provides faster and more memory efficient way of constructing component-wise acyclic subcomplex than Algorithm 4.1. The nodes of the resulting graph G are these simplices in $S_{max}(\mathcal{S})$ which are not in the acyclic subset. The algorithm uses the general graph functions AddToGraph and RemoveFromGraph, respectively adding a simplex to or removing a simplex from

---

Algorithm 4.2. AccIG(MaximalSimplexList $S_{max}(S)$, VertexHash H)

1: $V := \emptyset$; $E := \emptyset$; $\mathcal{A} := \emptyset$; $\mathcal{Q} :=$ EmptyQueue;
2: **for all** Simplex $P \in S_{max}(S)$ **do**
3:     **if** $P \notin V$ **and** $P \notin \mathcal{A}$ **then**
4:         $\mathcal{A} := \mathcal{A} \cup \{P\}$;
5:         EnqNeighb$(P, H, \mathcal{Q})$;
6:         **while** $\mathcal{Q} \neq \emptyset$ **do**
7:             Simplex $current :=$ Dequeue$(\mathcal{Q})$;
8:             **if** AcyclicityTest$(\mathcal{A}, current) =$ true **then**
9:                 $\mathcal{A} := \mathcal{A} \cup \{current\}$;
10:                EnqNeighb$(current, H, \mathcal{Q})$;
11:                **if** $current \in V$ **then**
12:                    RemoveFromGraph$(current, V, E)$;
13:             **else if** $current \notin V$ **then**
14:                 AddToGraph$(current, V, E, H)$;
15:                 EnqNeighb$(current, H, \mathcal{Q})$;
16: **return** Graph(V, E), $\mathcal{A}$;

---

a given graph. It also uses function `EnqNeighb`, which enqueues all neighbors of given simplex that are not yet in the queue nor in the acyclic subset.

Theorem 4.2. Algorithm 4.2 stops and returns a component-wise acyclic complex $\mathcal{A}$ and the incidence graph $G = (V, E)$ whose nodes are the maximal simplices in $S_{max}(S) \setminus \mathcal{A}$.

Proof. A simplex may be added to $\mathcal{Q}$ only if it does not belong to the acyclic subset and its neighbor has been added to the incidence graph or the acyclic subset. Since each simplex can be added to the graph or the acyclic subset at most once, the algorithm stops.

We start building a new acyclic component of the set $\mathcal{A}$ by finding a simplex that has not been added yet neither to the graph nor to the acyclic subset. Thus, it is not a neighbor of any simplex already processed. It means it represents new connected component of S in which we can start build new acyclic subset $\mathcal{A}$. We extend it only by adding those maximal simplices that have acyclic intersection with $\mathcal{A}$. For every simplex in $\mathcal{Q}$ we either add it to the acyclic subset or to the incidence graph, which means that nodes of the created incidence graph are all maximal simplices of $S_{max}(S)$ that have not been added to $\mathcal{A}$. $\square$

5   Distributed Computations

In this section we show how the algorithms that compute a component-wise acyclic subcomplex for a given simplicial complex can be used in distributed computations. The idea is to divide the initial complex into small parts, then construct a component-wise acyclic subcomplex and an incidence graph for each part and finally combine the results into a component-wise acyclic subcomplex and incidence graph for the initial complex. However, we need to ensure that after combining the results from the individual computations the obtained space is component-wise acyclic, i.e. we do not make cycles while connecting the acyclic subsets from the different parts. Moreover, we need a way to connect individual incidence graphs into the incidence graph of the initial complex. The whole procedure is very technical and resembles what we do in Algorithm 4.1 but in a more global scale. Let us emphasize that distributed computations involve only the construction of the incidence graph and the acyclic subset for each part. After combining the results from the individual reductions we create one complex for which we can perform homology computations just like in the non-distributed case.

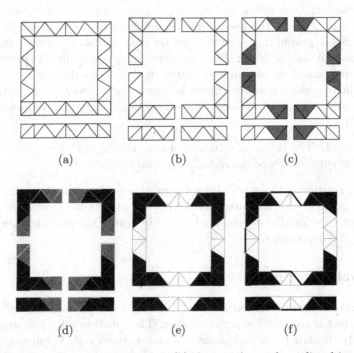

(a)   (b)   (c)

(d)   (e)   (f)

Fig. 1. (a) the initial simplicial complex, (b) the initial complex splitted into smaller, partial complexes, (c) the boundary simplices in the partial complexes, (d) the acyclic subsets in the partial complexes (black), (e) the combined results, (f) the acyclic subsets joined with a spanning forest (black)

The first step is to split the initial list of maximal simplices of S (Figure 1a) into lists $\mathcal{P}_i$, $i \in \{1, 2, ..., n\}$ in such a way that $\bigcup_i \mathcal{P}_i = S_{max}(S)$ (Figure 1b) and $\mathcal{P}_i \cap \mathcal{P}_j = \emptyset$ if $i \neq j$. For every $\mathcal{P}_i$ we define two sets: $BV_i := \bigcup_{i \neq j} \{S_0(\mathcal{P}_i) \cap S_0(\mathcal{P}_j)\}$ and $BS_i := \{Q \mid Q \in \mathcal{P}_i \wedge S_0(Q) \cap BV_i \neq \emptyset\}$. The elements of BS_i are referred to as the *boundary simplices* - simplices which have neighborhood contained in other packages. (Figure 1c). In the process of constructing the acyclic subset $\mathcal{A}_i$ for each $\mathcal{P}_i$ we consider only those simplices that are not boundary simplices (Figure 1d). To do so, we need to change a little Algorithms 4.1 and 4.2 so they include such restriction. We will not present them here, but it is easy for the reader to do such modification. In our example (Figure 1d) acyclic subset is constructed from all simplices that are not boundary simplices, but in general case this is not true. Computations of lists of both incidence graphs G_i and acyclic subsets $\mathcal{A}_i$ may be performed sequentially or in a distributed manner. In both cases we gain profits from lower memory usage, because list of simplices for which computations are performed are much smaller than the initial one. In the second case computations are performed much faster. Moreover, after constructing the acyclic subsets $\mathcal{A}_i$ we can discard all simplices contained in $\mathcal{A}_i$ from the incidence graph and construct a new acyclic subset which is the intersection of $\mathcal{A}_i$ with the lower dimensional faces of simplices which are left in the incidence graph. In the latter case, we save additional memory needed to store redundant simplices. Finally, after combining the results (Figure 1e) into one incidence graph we obtain a structure analogous to the one in Algorithm 4.1. We then create a spanning forest in which nodes are disjoint parts of the acyclic subset and edges are lists of one-dimensional simplices connecting them (Figure 1f).

Theorem 5.1. The family of simplices $\mathcal{A}$ constructed as above is a component-wise acyclic subcomplex of the initial simplicial complex $\mathcal{S}$.

Proof. By restricting the acyclic subset algorithm to these simplices that are not boundary simplices we are sure that the acyclic subsets in the respective parts do not create cycles after combining them. The rest of the proof is analogous to the proof of Theorem 4.1. $\square$

6   Acyclicity Tests

The `AcyclicityTest` function is a tool allowing to decide whether we can add a simplex to the constructed acyclic subset. The function takes two arguments: the already constructed acyclic subset $\mathcal{A}$ and a simplex P. We distinguish two types of acyclicity tests:

- a *full test* – it returns **true** if and only if $\mathcal{A} \cap cl(P)$ is acyclic
- a *partial test* – if it returns **true**, then $\mathcal{A} \cap cl(P)$ is acyclic but **false** on output denotes a failure to prove that $\mathcal{A} \cap cl(P)$ is acyclic.

Acyclicity tests in the setting of cubical sets, both full and partial, are proposed in [8].

The main limitation for quick acyclicity tests is the dimension of the complex. The full tests both in the cubical [8] and in the simplicial case are based on the idea of *tabulated configurations* for boundary elements. The number of configurations is 2^{3^d-1} for a d-dimensional cube and $2^{2^{d+1}}$ for a d-dimensional simplex. This makes the method prohibitive for $d > 3$ in the case of cubical sets [8] and for

$d > 4$ in the case of simplicial complexes [3]. The universal full test that works for every dimension is the computation of the homology of $\mathcal{A} \cap cl(P)$. However, this method is computationally very expensive. Therefore, in the dimensions where the tabulated configurations cannot be used, the use of quick partial tests is of interest. An acyclic subset algorithm based on partial acyclicity tests remains correct and often provides acyclic subsets which are not substantially smaller or even the same size as the algorithms based on full tests.

In the rest of this section we introduce few partial tests for the simplicial case. Given an acyclic subspace $\mathcal{A}$ and a d-dimensional simplex P we set $\mathcal{I} := \mathcal{A} \cap cl(P)$. The first test is based on the investigation of the maximal simplices of $\mathcal{I}$. It is straightforward to check that if the number of maximal simplices of dimension $d - 1$ in $\mathcal{I}$ is less than or equal to d and there are no maximal simplices of other dimensions in $\mathcal{I}$, then $\mathcal{I}$ is acyclic. This proves the following theorem.

Algorithm 6.1. AccTestCoDim1(Set $\mathcal{A}$, Simplex P)

1: $\mathcal{I} :=$ MaximalSimplices($\mathcal{A} \cap cl(P)$);
2: $d :=$Dim(P); $i := 0$;
3: **for all** Simplex $Q \in \mathcal{I}$ **do**
4:     **if** Dim(Q) $= d - 1$ **then**
5:         i++;
6:     **else**
7:         return false;
8: **if** $i > 0$ and $i <= d$ **then**
9:     return true;
10: **else**
11:     return false;

Theorem 6.1. Given a set $\mathcal{A}$ and simplex P on input, if Algorithm 6.1 returns **true**, then $\mathcal{A} \cap cl(P)$ is acyclic. However, **false** on output denotes a failure to decide whether $\mathcal{A} \cap cl(P)$ is acyclic.

Algorithm 6.2 tries to find a vertex of P which is a common face of all maximal simplices in $\mathcal{I}$. If it is able to do so, it means that $\mathcal{I}$ forms a topology of a star and therefore is acyclic. Analogous theorem as for Algorithm 6.1 can be stated for Algorithm 6.2.

Two more partial tests will be introduced without presenting suitable algorithms and theorems. First of them uses the list of maximal simplices $\mathcal{I}$ for construction of an acyclic subcomplex $\mathcal{I}'$ of $\mathcal{I}$. The whole procedure is exactly the same as presented in this paper. For testing acyclicity it uses itself recursively. At the bottom of recursion we only need to determine if the intersection of two one dimensional simplices is acyclic. In fact it is true if and only if simplices share common vertex, which is trivial to check. If constructed acyclic subcomplex $\mathcal{I}'$ is the same as the initial complex $\mathcal{I}$, then $\mathcal{I}$ is acyclic.

The last test constructs simplicial complex from $\mathcal{I}$ and performs coreductions [7] on it. If the resulted complex is fully reduced, it means that $\mathcal{I}$ is acyclic.

Algorithm 6.2. AccTestStar(Set $\mathcal{A}$, Simplex P)

1: $\mathcal{I} :=$ MaximalSimplices($\mathcal{A} \cap cl(P)$);
2: **for all** Vertex $v \in P$ **do**
3:　　ok := true;
4:　　**for all** Simplex $Q \in \mathcal{I}$ **do**
5:　　　　**if** $v \subsetneq Q$ **then**
6:　　　　　　ok := false;
7:　　　　　　break;
8:　　**if** ok **then**
9:　　　　return true;
10: return false;

---

7 Numerical Experiments

All algorithms presented in this paper have been implemented in C++. The code will be available as a part of RedHom [11] library. To provide a communication between processes during distributed computation MPI [4] was used. Both local and distributed approaches were compared with the coreduction homology algorithm [7], denoted in the following table by CoRed. AccIG and AccST are the algorithms introduced in Section 4. DAccIG and DAccST denote the outcome of distributed computations using AccIG and AccST algorithms respectively for a local construction of an acyclic subspace. Column size denotes the number of maximal simplices used as input. Value in column s is the total time in seconds needed for building the incidence graph, performing reductions (which could be either removal of acyclic subset or coreductions [7]), creating the simplicial complex from the list of maximal simplices and computing Betti numbers for the complex [5]. Column MB contains the total amount of memory in megabytes needed for performing computations. Distributed computations were performed on 4 nodes (1 master and 3 slaves) simultaneously and values in this case denotes the maximum running time in seconds over all nodes and maximum amount of memory that single node needs. Computing generators after reduction of acyclic subset is still an open problem.

Space name	Size	CoRed		AccIG		AccST		DAccIG		AccST	
		s	MB	s	MB	s	MB	s	MB	s	MB
Bjorner	3079k	174	2330	203	1035	375	2968	154	568	229	1008
Dunce Hat	4758k	273	3603	327	1647	598	4525	224	821	390	1527
Proj. Plane	2799k	158	2180	189	943	323	2638	143	437	190	900

The second table contains comparison of efficiency of acyclicity test algorithms presented in Section 6. Algorithm denoted as Tab is acyclicity test that uses tabulated configurations. CoDim1 and Star are respectively Algorithms 6.1 and 6.2. Rec is recursive test, Hom is full test that uses homology computations and Cored is test based on coreductions. For each algorithm column s denotes total

Space name	Size	Tab		CoDim1		Star		Rec		Hom		Cored	
		s	#	s	#	s	#	s	#	s	#	s	#
Bjorner	513216	33	513215	49	513215	34	513215	39	513215	166	513215	124	513215
Dunce Hat	793152	53	789279	73	789060	50	789279	55	789279	253	789279	192	789279
Proj. Plane	466560	29	464511	42	464609	31	464511	32	464511	150	464511	111	464511

running time in seconds needed for performing computations, just as described above, using AccIG algorithm and selected acyclicity test. Column # denotes number of maximal simplices in constructed acyclic subcomplex.

Acknowledgments. P.B. and M.M. are partially supported by Polish MNSzW, Grant N N201 419639. P.D. is partially supported by grant Nr IP 2010 046370.

References

1. Cormen, T.H., Leiserson, C.E., Rivest, R.L.: Introduction to Algorithms. MIT Press and McGraw-Hill (1990)
2. Dłotko, P., Specogna, R.: Efficient Cohomology Computation for Electromagnetic Modeling. Computer Modeling in Engineering & Sciences 60(3), 247–277 (2010)
3. Dłotko, P.: Acyclic configurations for boundary elements of 3 and 4 dimensional simplices, http://www.ii.uj.edu.pl/~dlotko/accconf.html
4. Gropp, W., Lusk, E., Skjellum, A.: Using MPI: Portable Parallel Programming with the Message-Passing Interface. MIT Press (1990)
5. Kaczynski, T., Mischaikow, K., Mrozek, M.: Computational homology, Appl. Math. Sci., vol. 157. Springer, New York (2004)
6. Kaczynski, T., Mrozek, M., Ślusarek, M.: Homology computation by reduction of chain complexes. Computers and Math. Appl. 35, 59–70 (1998)
7. Mrozek, M., Batko, B.: Coreduction Homology Algorithm. Discrete and Computational Geometry 41, 96–118 (2009)
8. Mrozek, M., Pilarczyk, P., Żelazna, N.: Homology algorithm based on acyclic subspace. Computers and Mathematics with Applications 55, 2395–2412 (2008)
9. Munkres, J.R.: Elements of Algebraic Topology. Perseus Publishing, Cambridge (1984)
10. Computer Assisted Proofs in Dynamics, http://capd.wsb-nlu.edu.pl
11. The RedHom homology algorithms library, http://redhom.ii.uj.edu.pl

Multi-scale Approximation of the Matching Distance for Shape Retrieval

Andrea Cerri[1], Barbara Di Fabio[1], and Filippo Medri[2]

[1] ARCES, Università di Bologna, Italia
{andrea.cerri2,barbara.difabio}@unibo.it
[2] Dipartimento di Scienze dell'Informazione, Università di Bologna, Italia
filippo.medri@gmail.com

Abstract. This paper deals with the concepts of persistence diagrams and matching distance. They are two of the main ingredients of *Topological Persistence*, which has proven to be a promising framework for shape comparison. Persistence diagrams are descriptors providing a signature of the shapes under study, while the matching distance is a metric to compare them. One drawback in the application of these tools is the computational costs for the evaluation of the matching distance. The aim of the present paper is to introduce a new framework for the approximation of the matching distance, which does not affect the reliability of the entire approach in comparing shapes, and extremely reduces computational costs. This is shown through experiments on 3D-models.

Keywords: Persistence diagram, shape analysis, dissimilarity criterion.

1   Introduction

Interpreting and comparing shapes are challenging issues in computer vision, computer graphics and pattern recognition [11,12]. Topological Persistence – including Persistent Homology [9] and Size Theory [1,10] – has proven to be a successful comparison/retrieval/classification (hereafter CRC) scheme.

In a nutshell, the basic idea for dealing with the CRC task is to define a measure of the (dis)similarity between the shapes in a given database. This can be done by extracting a battery of shape descriptors – the so-called *persistence diagrams* – from each element in the database, capturing meaningful shape properties. Thus, the problem of assessing the (dis)similarity between two shapes can be recast into the one of comparing the associated persistence diagrams according to the *matching* (or *bottleneck*) *distance*, a proven stable distance between these descriptors. This process defines a metric over the database, that can be used for CRC purposes. In general, a given persistence diagram may come from different shapes: This can be interpreted as an equivalence with respect to the properties captured by that descriptor.

Such an approach has been successfully used in a number of concrete problems concerning shape comparison and retrieval [4,5,8]. However, defining a (dis)similarity metric in the case of large databases can lead to considerable

M. Ferri et al. (Eds.): CTIC 2012, LNCS 7309, pp. 128–138, 2012.

computational costs. The bottleneck in this procedure can be identified in the evaluation of the matching distance.

The Contribution of the Paper. Reducing the computational costs in defining a (dis)similarity metric within a database of shapes is definitely a desirable target: This would enable us to further improve the persistence CRC framework and apply it to a wider class of concrete problems. The present paper aims to illustrate an idea to achieve this goal, ranging from a theoretical formalization of the proposed strategy to its validation through an experimental study. We introduce a multi-scale construction of our matching distance-based (dis)similarity metric. Our procedure is based on a "dissimilarity criterion" which is formalized in Theorem 1. Experiments on 3D-models show that, using our idea, it is possible not to affect the reliability of the entire approach in comparing shapes, extremely reducing the computational costs.

2   Preliminaries

In persistence, the shape of an object is usually studied by choosing a topological space X to represent it, and a function $\varphi : X \to \mathbb{R}$, called a *filtering (or measuring) function*, to define a family of subspaces $X_u = \varphi^{-1}((-\infty, u])$, $u \in \mathbb{R}$, nested by inclusion, i.e. a filtration of X. Applying homology to the filtration allows us to study how topological features vary in passing from a set of the filtration into a larger one, and to rank topological features with bounded lifetime by importance, according to the length of their life. The basic assumption is that the longer a feature survives, the more meaningful or coarse the feature is for shape description. Vice-versa, noise and shape details are characterized by a shorter life. For further details we refer to [1,9].

The filtration $\{X_u\}_{u \in \mathbb{R}}$ is used to define persistent homology groups as follows. Given $u \leq v \in \mathbb{R}$, we consider the inclusion of X_u into X_v. This inclusion induces a homomorphism of homology groups $H_k(X_u) \to H_k(X_v)$ for every $k \in \mathbb{Z}$. Its image consists of the k-homology classes that live at least from $H_k(X_u)$ to $H_k(X_v)$ and is called the *kth persistent homology group of (X, φ) at (u, v)*. When this group is finitely generated, we denote by $\beta_k^{u,v}(X, \varphi)$ its rank.

A simple and compact description of persistent homology groups of (X, φ) is provided by the so-called persistence diagrams, i.e. multisets of points whose abscissa and ordinate are, respectively, the level at which a new k-homology class is created and the level at which it is annihilated through the filtration.

We use the following notation: $\Delta^+ = \{(u, v) \in \mathbb{R}^2 : u < v\}$, $\Delta = \{(u, v) \in \mathbb{R}^2 : u = v\}$, and $\overline{\Delta^+} = \Delta^+ \cup \Delta$.

Definition 1 (Multiplicity). *Let $k \in \mathbb{Z}$ and $(u, v) \in \Delta^+$. The multiplicity $\mu_k(u, v)$ of (u, v) is the finite non-negative number defined by*

$$\lim_{\varepsilon \to 0^+} \left(\beta_k^{u+\varepsilon, v-\varepsilon}(X, \varphi) - \beta_k^{u-\varepsilon, v-\varepsilon}(X, \varphi) - \beta_k^{u+\varepsilon, v+\varepsilon}(X, \varphi) + \beta_k^{u-\varepsilon, v+\varepsilon}(X, \varphi) \right).$$

Definition 2 (Persistence Diagram). *The persistence diagram $D_k(X, \varphi)$ is the multiset of all points $(u, v) \in \Delta^+$ such that $\mu_k(u, v) > 0$, counted with their multiplicity, union the points of Δ, counted with infinite multiplicity.*

We will call *proper points* the points of a persistence diagram lying on Δ^+.

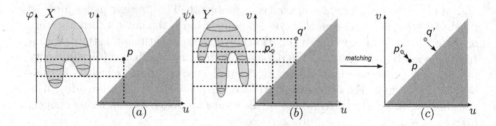

Fig. 1. (*a*) The height function φ on the space X, and the associated persistence diagram $D_0(X, \varphi)$. (*b*) The height function ψ on the space Y, and the associated persistence diagram $D_0(Y, \psi)$. (*c*) The matching between $D_0(X, \varphi)$ and $D_0(Y, \psi)$ realizing their matching distance.

Figures 1 (*a*) − (*b*) show two examples of persistence diagrams for $k = 0$. For instance, in Figure 1 (*a*) a surface $X \subset \mathbb{R}^3$ is filtered by the height function φ. The sole proper point of $D_0(X, \varphi)$ is p. Its abscissa corresponds to the level at which a new connected component is born, while its ordinate identifies the level at which this connected component merges with the existing one. To see, for instance, that $\mu_0(p) = 1$, letting $p = (\bar{u}, \bar{v})$, it is sufficient to observe that, for every $\varepsilon > 0$ sufficiently small, it holds that $\beta_0^{\bar{u}+\varepsilon, \bar{v}-\varepsilon}(X, \varphi) = 2$, $\beta_0^{\bar{u}-\varepsilon, \bar{v}-\varepsilon}(X, \varphi) = \beta_0^{\bar{u}+\varepsilon, \bar{v}+\varepsilon}(X, \varphi) = \beta_0^{\bar{u}-\varepsilon, \bar{v}+\varepsilon}(X, \varphi) = 1$, and apply Definition 1.

The matching distance between two persistence diagrams measures the cost of finding a correspondence between their points. In doing this, the cost of taking a point p to a point p' is measured as the minimum between the cost of moving one point onto the other and the cost of moving both points onto the diagonal, see Figure 1 (*c*) for an example. In particular, the matching of a proper point p with a point of Δ can be interpreted as the destruction of the point p. Formally:

Definition 3 (Matching Distance). *Let D_k^1, D_k^2 be two persistence diagrams. The matching distance $d_{match}\left(D_k^1, D_k^2\right)$ is defined as*

$$d_{match}(D_k^1, D_k^2) = \min_{\sigma} \max_{p \in D_k^1} d(p, \sigma(p)),$$

where σ varies among all the bijections between D_k^1 and D_k^2 and

$$d\left((u, v), (u', v')\right) = \min\left\{\max\left\{|u - u'|, |v - v'|\right\}, \max\left\{\frac{v - u}{2}, \frac{v' - u'}{2}\right\}\right\} \quad (1)$$

for every $(u, v), (u', v') \in \overline{\Delta^+}$.

The importance of the matching distance in persistence is based on the fact that persistence diagrams are robust with respect to it. Roughly, small changing in a given filtering function (w.r.t. the max-norm) produces just a small changing in the associated persistence diagram w.r.t. the matching distance [7,9].

Remark 1. From Definition 3 it follows that $d_{match}(D_k^1, D_k^2) \leq (V - U)/2$, with $U = \min\limits_{(u,v)\in L} u$, $V = \max\limits_{(u,v)\in L} v$ and $L = D_k^1 \cup D_k^2$. Indeed, $(V - U)/2$ upper bounds the cost of the bijection between D_k^1 and D_k^2, taking all the points of L onto Δ. Since d_{match} is realized by the cheapest bijection between D_k^1 and D_k^2, we have the claim.

This result will be useful later.

3   Theoretical Setting and Results

Computationally, evaluating the matching distance between two persistence diagrams takes $O(h^{2.5})$ [6], being h the total amount of their proper points.

As stressed before, in CRC applications involving large databases, computing the matching distance for any possible shape comparison can imply a high computational cost. In fact, noisy or detailed shape models can produce persistence diagrams with a large number of proper points. Our goal is to reduce this computational complexity by contenting, at first, of a rough estimation of the metric induced by the matching distance over a database, to be possibly refined whenever it is not sufficient to distinguish between different shapes.

The key point here is the observation that, in most cases, realizing that two shapes are very dissimilar does not require to compute the *exact* matching distance between the associated persistence diagrams. Deciding, e.g., whether an elephant is different from an ant requires only a first glance at the two animals. In our framework, such a "first glance" could be equivalent to a rough estimation of the matching distance – and hence faster than its exact computation – between the persistence diagrams associated with the "elephant shape" and the "ant shape", respectively. On the contrary, a different level of accuracy could be necessary to distinguish, e.g., the "wolf shape" from the "German shepherd shape". This would lead to a sharper estimation of the matching distance between the associated persistence diagrams, possibly to its actual computation.

In light of these considerations, we propose a multi-scale construction of our matching distance-based (dis)similarity metric.

Let D_k be a persistence diagram. For every $p = (u, v) \in \Delta^+$ and every $\delta > 0$, let $\mathcal{Q}_\delta(p)$ be the open square centered at p of side equal to 2δ, and let us denote by $\sharp(\mathcal{Q}_\delta(p), D_k)$ the number of points of D_k contained in $\mathcal{Q}_\delta(p)$.

Theorem 1 (Dissimilarity Criterion). *Let D_k^1, D_k^2 be two persistence diagrams for which a point $p = (u, v) \in \Delta^+$ and two real numbers $\delta, \varepsilon > 0$ exist, such that $\mathcal{Q}_{\delta+\varepsilon}(p) \subset \Delta^+$ and $\sharp(\mathcal{Q}_\delta(p), D_k^1) - \sharp(\mathcal{Q}_{\delta+\varepsilon}(p), D_k^2) > 0$. Then $d_{match}(D_k^1, D_k^2) \geq \varepsilon$.*

Proof. Since $\sharp(\mathcal{Q}_\delta(p), D_k^1) > \sharp(\mathcal{Q}_{\delta+\varepsilon}(p), D_k^2)$, for every bijection $\sigma : D_k^1 \rightarrow D_k^2$ there exists at least one proper point $\bar{q} = (\bar{u}, \bar{v}) \in D_k^1$ such that $\bar{q} \in \mathcal{Q}_\delta(p)$ and $\sigma(\bar{q}) = \bar{q}' = (\bar{u}', \bar{v}') \in D_k^2$, with $\bar{q}' \notin \mathcal{Q}_{\delta+\varepsilon}(p)$. Then, from (1) it holds that

$$d(\bar{q}, \bar{q}') \geq \min\left\{\varepsilon, \max\left\{\frac{\bar{v} - \bar{u}}{2}, \frac{\bar{v}' - \bar{u}'}{2}\right\}\right\} \geq \min\left\{\varepsilon, \frac{\bar{v} - \bar{u}}{2}\right\} = \varepsilon. \quad (2)$$

Indeed, in (2), the first inequality holds because both $|\bar{u} - \bar{u}'|$ and $|\bar{v} - \bar{v}'|$ are not smaller than the difference between the semi-sides of $\mathcal{Q}_\delta(p)$ and $\mathcal{Q}_{\delta+\varepsilon}(p)$; the second inequality is obvious; the equality follows from both the facts that $\bar{v} - \bar{u} > (v - \delta) - (u + \delta)$, being $(\bar{u}, \bar{v}) \in \mathcal{Q}_\delta(p)$ and $(u + \delta, v - \delta) \in \Delta^+$ the bottom right vertex of $\mathcal{Q}_\delta(p)$, and $(v - \delta - \varepsilon) - (u + \delta + \varepsilon) \geq 0$, i.e. $(v - \delta) - (u + \delta) \geq 2\varepsilon$, being $(u + \delta + \varepsilon, v - \delta - \varepsilon) \in \overline{\Delta^+}$ the bottom right vertex of $\mathcal{Q}_{\delta+\varepsilon}(p)$. Hence $\max_{q \in D_k^1} d(q, \sigma(q)) \geq \varepsilon$ for every bijection σ and, by Definition 3, the claim is proved.

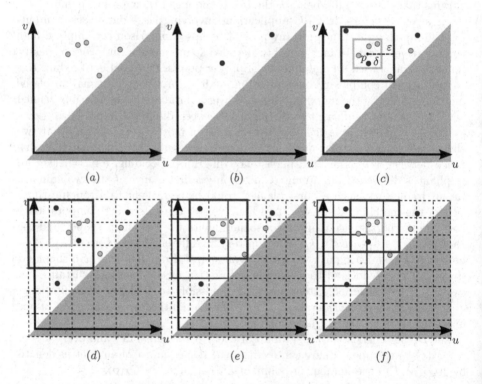

Fig. 2. $(a) - (b)$ Two persistence diagrams D_k^1 and D_k^2. (c) The overlapping of D_k^1 and D_k^2, and the two squares $\mathcal{Q}_\delta(p)$ and $\mathcal{Q}_{\delta+\varepsilon}(p)$ for a certain $p \in \Delta^+$. $(d) - (f)$ Algorithm 1 in action: three steps are necessary to find squares in which Theorem 1 holds.

Figures 2 $(a) - (c)$ show an example of Theorem 1 in action. Figures 2 $(a) - (b)$ represent two persistence diagrams, say D_k^1 and D_k^2, respectively. In Figure 2 (c) the two multisets of points are overlapped, and the two squares $\mathcal{Q}_\delta(p)$ and $\mathcal{Q}_{\delta+\varepsilon}(p)$ are depicted. As can be seen, it holds that $\sharp(\mathcal{Q}_\delta(p), D_k^1) - \sharp(\mathcal{Q}_{\delta+\varepsilon}(p), D_k^2) = 1$. Hence, by Theorem 1 we get that surely $d_{match}(D_k^1, D_k^2) \geq \varepsilon$.

The issue here is to find a suitable way to apply Theorem 1, so to improve our CRC framework. This is what the following Algorithm 1 is thought for.

Algorithm 1 takes as input the lists A and B of proper points of two persistence diagrams, and a parameter Exp. It runs a number of iterations equal to $\lfloor h^{Exp} \rfloor$ where $h = |A| + |B|$ is the sum of the number of points of A and B, and Exp is an arbitrary positive rational number. During each iteration, a finer grid is created on a triangular region $T \subset \overline{\Delta^+}$ with vertices $(U - \varepsilon, U - \varepsilon), (U - \varepsilon, V + \varepsilon), (V + \varepsilon, V + \varepsilon)$, being U and V as in Remark 1, containing all the points belonging to A and B. In particular, at each iteration n, the algorithm produces $n(n+1)/2$ small squares with side equal to $(n+5)th$ part of the side of T. It then evaluates Theorem 1 on each small square compared with the square having its same center and side three times greater. The algorithm returns the maximum value for which Theorem 1 holds. Algorithm 1 makes use of two different subroutines: $Matrix(i, j)$ which simply generates a two dimensional matrix $0_{i \times j}$ and $CountPoints(S, p, q)$ (Algorithm 2) whose output is the sum of the entries of the 3×3 submatrix $S[p-1, p, p+1; q-1, q, q+1]$ which is nothing more than the number of points of the largest square into which we are going to evaluate the theorem. Algorithm 3 gives as output the actual or the approximated distance between two persistence diagrams. An example of Algorithm 1 in action is shown in Figures 2 $(d) - (f)$.

Algorithm 1. MatchDistGridApprox(A, B, Exp)

1: $N \Leftarrow \lfloor (|A| + |B|)^{Exp} \rfloor$	16: $\quad\quad qB(i, j) \Leftarrow qB(i, j) + 1$
2: $Res \Leftarrow 0$	17: $\quad$ **end for**
3: $\varepsilon \Leftarrow (V - U)/10$	18: $\quad$ **for** $p = 2$ **to** $(t - 1)$ **do**
4: $Side \Leftarrow V - U + 2\varepsilon$	19: $\quad\quad$ **for** $q = (p + 3)$ **to** $(t - 1)$ **do**
5: **for** $n = 1$ **to** N **do**	20: $\quad\quad\quad QA \Leftarrow$ CountPoints(qA, p, q)
6: $\quad t \Leftarrow 5 + n$	21: $\quad\quad\quad QB \Leftarrow$ CountPoints(qB, p, q)
7: $\quad sSide \Leftarrow Side/t$	22: $\quad\quad\quad r1 \Leftarrow (QA < qB(p, q))$
8: $\quad qA \Leftarrow$ Matrix(t, t)	23: $\quad\quad\quad r2 \Leftarrow (QB < qA(p, q))$
9: $\quad qB \Leftarrow$ Matrix(t, t)	24: $\quad\quad\quad$ **if** $(r1$ **or** $r2)$ **and** $(Res < sSide)$ **then**
10: $\quad$ **for all** $a \in A$ **do**	25: $\quad\quad\quad\quad Res \Leftarrow sSide$
11: $\quad\quad (i, j) \Leftarrow \lceil (a - U + \varepsilon)/sSide \rceil$	26: $\quad\quad\quad$ **end if**
12: $\quad\quad qA(i, j) \Leftarrow qA(i, j) + 1$	27: $\quad\quad$ **end for**
13: $\quad$ **end for**	28: $\quad$ **end for**
14: $\quad$ **for all** $b \in B$ **do**	29: **end for**
15: $\quad\quad (i, j) \Leftarrow \lceil (b - U + \varepsilon)/sSide \rceil$	30: **return** Res

---

The computational complexity C of Algorithm 1 can be formalized as

$$C(h, Exp) = c_1 + \sum_{n=1}^{h^{Exp}} \left(c_2 + 2c_3(n + 5)^2 + c_4 \cdot h + \sum_{p=2}^{n+4} \sum_{q=p+3}^{n+4} c_5 \right),$$

with $c_4 \cdot h$ the cost of lines $10 - 17$, $c_3(n + 5)^2$ the cost of lines $8 - 9$, c_3 and c_4 being constants as well as c_1 (lines $1 - 4$), c_2 (lines $6 - 7$) and c_5 (lines $20 - 30$).

Making some simple mathematical manipulations we obtain that

$$C(h, Exp) = c_1 + h^{Exp}(c_2 + c_4 \cdot h) + 2c_3 \cdot \sum_{n=1}^{h^{Exp}}(n + 5)^2 + \sum_{n=1}^{h^{Exp}} \sum_{p=1}^{n+3} \sum_{q=1}^{n-p+1} c_5.$$

Now, by counting the total number of squares on which the theorem is evaluated on a run of the algorithm, which is

$$\sum_{n=1}^{h^{Exp}} \sum_{p=1}^{n+3} \sum_{q=1}^{n-p+1} 1 = \sum_{n=1}^{h^{Exp}} \sum_{p=1}^{n+3}(n-p+1) = \sum_{n=1}^{h^{Exp}} \frac{n(n+1)}{2} = \frac{h^{3Exp} + 3h^{2Exp} + 2h^{Exp}}{6},$$

we can conclude that the computational complexity of Algorithm 1 is $O(h^{3Exp})$. Hence, by choosing $Exp \leq \frac{2.5}{3}$ we can ensure that Algorithm 1 has a computational complexity asymptotically lower than the one we would have by calculating the matching distance.

Algorithm 2. CountPoints(S, p, q)	**Algorithm 3.** MetricApprox($A, B, Exp, thresh$)
1: **for** $i = (p-1)$ to $(p+1)$ **do**	1: $Res =$ MatchDistGridApprox(A, B, Exp)
2: **for** $j = (q-1)$ to $(q+1)$ **do**	2: **if** $Res > thresh$ **then**
3: $Res \Leftarrow Res + S(i, j)$	3: $Val = [(V - U)/2 + Res]/2$
4: **end for**	4: **else**
5: **end for**	5: $Val = d_{match}(A, B)$
6: **return** $Result$	6: **return** Val

4   Experimental Results

Our goal is to validate the theoretical framework introduced in the previous section. Through some experiments on persistence diagrams for 0th homology degree (a.k.a. *formal series* [10]), associated with 3D-models represented by triangle meshes, we will prove that our algorithm allows us to reduce the computational complexity in defining a matching distance-based metric over a given database, without greatly affecting the goodness of results (in terms of database classification).

To test the proposed framework we considered a database of 228 3D-surface mesh models introduced in [2]. The database is divided into 12 classes, each containing 19 elements obtained as follows: A null model taken from the Non Rigid World Benchmark [3] is considered together with six non-rigid transformations applied to it at three different strength levels. An example of the transformations and their strength levels is given in Table 1. To define the considered filtering functions, we proceeded as follows: For each triangle mesh M of vertices

$\{P_1, \ldots, P_n\}$, the center of mass B is computed, and the model is normalized to be contained in a unit sphere. Further, a vector $\boldsymbol{w}$ is defined as

$$\boldsymbol{w} = \frac{\sum_{i=1}^{n}(P_i - B)\|P_i - B\|}{\sum_{i=1}^{n}\|P_i - B\|^2}.$$

Three filtering functions $\varphi_1, \varphi_2, \varphi_3$ are computed on the vertices of M: φ_1 is the distance from the line parallel to $\boldsymbol{w}$ and passing through B, φ_2 is the distance from the plane orthogonal to $\boldsymbol{w}$ and passing through B, and φ_3 is the distance from B. The values of φ_1, φ_2 and φ_3 are then normalized so that they range in the interval $[0, 1]$. These filtering functions are translation and rotation invariant, as well as scale invariant because of a priori normalization of the models. Moreover, the considered models are sufficiently generic (no point-symmetries occur etc...) to ensure that the vector $\boldsymbol{w}$ is well-defined over the all database, as well as its orientation is stable.

Taking a filtering function φ, we can now induce a metric over our database by computing the matching distances $d_{ij}^{\varphi} = d_{match}(D_0(M_i, \varphi), D_0(M_j, \varphi))$ for every $i, j = 1, \ldots, 228$. To approximate such a metric, we applied Algorithm 1 to get a lower bound for each d_{ij}^{φ}, say Res_{ij}^{φ}. This procedure is controlled by a threshold, $thresh^{\varphi}$, obtained as follows: For every class in the database, 4 elements are (randomly) selected, and an average of the matching distances on this small subset is evaluated. The final value of $thresh^{\varphi}$ is then the average over all the classes in the database. In this perspective, the value $thresh^{\varphi}$ represents the average matching distance between two elements of the same class.

Now, if $Res_{ij}^{\varphi} > thresh^{\varphi}$, then we can assume that the shapes of M_i and M_j are quite dissimilar (compared w.r.t. φ) and therefore it is sufficient to have just an estimation of d_{ij}^{φ}: We opted for $((V - U)/2 + Res_{ij}^{\varphi})/2$, with V and U taken according to Remark 1. If $Res_{ij}^{\varphi} \leq thresh^{\varphi}$, then the exact value of d_{ij}^{φ} is computed. The overall process is described in Algorithm 3.

Table 2 (first column) shows the average precision/recall (PR) graphs induced by φ_1, φ_2 and φ_3, respectively, when considering the computation of the matching distances on the whole database and on some subparts of it after running Algorithm 1, with Exp set at two different values. As can be seen, our approximation strategy does not affect so much the PR performances even in the displayed worst case (filtering function φ_2).

Table 2 (second column) gives a more general overview of the obtained results. From top to bottom, each graph shows the reduction in the computational costs – in terms of the percentage of computed matching distances used to build the metric approximations – and an evaluation of the PR performances according to the chosen values of Exp, for the filtering functions φ_1, φ_2 and φ_3, respectively. In particular, for a given value of Exp the evaluation of results is expressed as the average $L_1^{\bullet}$-distance between the PR graph associated to that value Exp and the one obtained by computing all the matching distances between the elements in the database. The "critical Exp" depicted in all plots represents the value of Exp such that the cost of applying Algorithm 1 equals the one of computing the matching distance between two persistence diagrams.

Table 1. The null model "Centaur0" and the 3rd strength level for each deformation

Table 2. First column: PR graphs related to $\varphi_1, \varphi_2, \varphi_3$ computing d_{match} on the whole database (black), and on subparts of it (PR approx) by virtue of Algorithm 1 for two different values of Exp (shaded); Second column: varying Exp, how the percentage of d_{match} computed and the distance between PR graph and PR approx vary

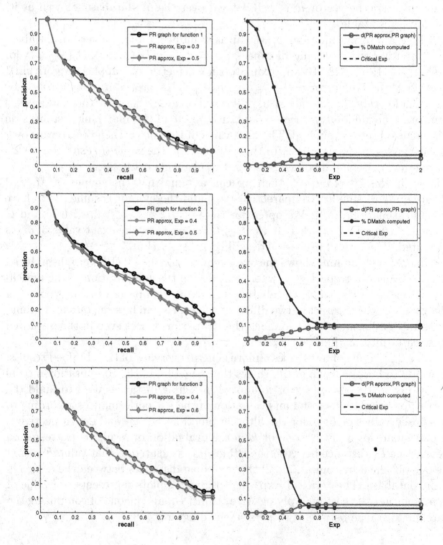

As our plots show, it is possible to greatly reduce the computational costs by approximating the matching distance-based metric over the database, obtaining PR graphs which are quite close to the best possible.

5 Conclusions

In this paper we introduced a multi-scale strategy to evaluate a (dis)similarity metric on a database of shapes – to be used for CRC purposes – using the concepts of persistence diagrams and matching distance. The proposed framework has been validated through experiments on 3D models represented by triangle meshes: The obtained results show that it is possible to provide an approximation of the metric induced by the matching distance between persistence diagrams without compromising the goodness of results – in terms of retrieval performance – and greatly reducing the computational costs coming from the exact evaluation of the matching distance.

For the next future we plan to generalize Algorithm 1 in such a way that the lower bound provided by Theorem 1 could be better exploited. We plan to do this by randomly generating the largest squares in the triangular area T, considered in Algorithm 1, allowing in this way partial covering of T and squares overlapping, and making that the smallest squares vary inside the wider ones. The expected result is to produce statistically better estimates of the matching distance lower bound through the use of a more flexible tool not stuck on a fixed tessellation like the one produced by Algorithm 1.

Acknowledgments. The authors wish to thank P. Frosini for suggesting the problem. However, the authors are solely responsible for any possible errors.

References

1. Biasotti, S., De Floriani, L., Falcidieno, B., Frosini, P., Giorgi, D., Landi, C., Papaleo, L., Spagnuolo, M.: Describing shapes by geometrical-topological properties of real functions. ACM Comput. Surv. 40(4), 1–87 (2008)
2. Biasotti, S., Cerri, A., Frosini, P., Giorgi, D.: A new algorithm for computing the 2-dimensional matching distance between size functions. Pattern Recognition Letters 32(14), 1735–1746 (2011)
3. Bronstein, A., Bronstein, M., Kimmel, R.: Numerical Geometry of Non-Rigid Shapes, 1st edn. Springer Publishing Company, Incorporated (2008)
4. Carlsson, G., Zomorodian, A., Collins, A., Guibas, L.J.: Persistence barcodes for shapes. IJSM 11(2), 149–187 (2005)
5. Chazal, F., Cohen-Steiner, D., Guibas, L.J., Mémoli, F., Oudot, S.: Gromov-Hausdorff stable signatures for shapes using persistence. Computer Graphics Forum 28(5), 1393–1403 (2009)
6. d'Amico, M., Frosini, P., Landi, C.: Using matching distance in size theory: A survey. Int. J. Imag. Syst. Tech. 16(5), 154–161 (2006)
7. d'Amico, M., Frosini, P., Landi, C.: Natural pseudo-distance and optimal matching between reduced size functions. Acta. Appl. Math. 109, 527–554 (2010)

8. Di Fabio, B., Landi, C., Medri, F.: Recognition of Occluded Shapes Using Size Functions. In: Foggia, P., Sansone, C., Vento, M. (eds.) ICIAP 2009. LNCS, vol. 5716, pp. 642–651. Springer, Heidelberg (2009)
9. Edelsbrunner, H., Harer, J.: Computational Topology: An Introduction. American Mathematical Society (2009)
10. Frosini, P., Landi, C.: Size theory as a topological tool for computer vision. Pattern Recogn. and Image Anal. 9, 596–603 (1999)
11. Smeulders, A.W.M., Worring, M., Santini, S., Gupta, A., Jain, R.: Content-based image retrieval at the end of the early years. IEEE Trans. PAMI 22(12) (2000)
12. Tangelder, J.W.H., Veltkamp, R.C.: A survey of content-based 3D shape retrieval methods. Multimedia Tools and Applications 39(3), 441–471 (2008)

Persistent Homology for 3D Reconstruction Evaluation

Antonio Gutierrez[1], David Monaghan[2],
María José Jiménez[1], and Noel E. O'Connor[2]

[1] Applied Math Department, School of Computer Engineering,
University of Seville,
Campus Reina Mercedes, 41012 Sevilla, Spain
majiro@us.es
[2] CLARITY: Centre for Sensor Web Technologies,
Dublin City University, Ireland
{david.monaghan,noel.oconnor}@dcu.ie

Abstract. Space or voxel carving is a non-invasive technique that is used to produce a 3D volume and can be used in particular for the reconstruction of a 3D human model from images captured from a set of cameras placed around the subject. In [1], the authors present a technique to quantitatively evaluate spatially carved volumetric representations of humans using a synthetic dataset of typical sports motion in a tennis court scenario, with regard to the number of cameras used. In this paper, we compute persistent homology over the sequence of chain complexes obtained from the 3D outcomes with increasing number of cameras. This allows us to analyze the topological evolution of the reconstruction process, something which as far as we are aware has not been investigated to date.

Keywords: voxel carving, volume reconstruction, persistent homology, evaluation.

1 Introduction

Homology is topologically invariant, meaning it is a property of an object that does not change under continuous (elastic) transformations of the object. Roughly speaking, homology characterizes "holes" in any dimension (e.g. connected components, tunnels and cavities in a 3D space). Homology computation can be carried out over a combinatorial structure called *cell complex*, which is built up by basic elements (*cells*) of different dimensions (vertices, edges, faces, etc.). One can take advantage of the combinatorial nature of a digital image (as a set of voxels) to compute homology by taking as input the (algebraic) cubical complex associated to the image. *Persistent homology* studies homology classes and their life-times (persistence) in the belief that significant topological attributes must have a long life-time in a filtration (an increasing nested sequence of subcomplexes). In this paper, we compute persistent homology via the Incremental and Decremental Algorithms for computing AT-models (see [5]), which allow to combine an incremental with a decremental technique in the case of a non-increasing

M. Ferri et al. (Eds.): CTIC 2012, LNCS 7309, pp. 139–147, 2012.

filtration, that is, a sequence of subcomplexes. In the following Section, we describe the context in which we apply persistent homology computation. Section 3 is devoted to recall basic tools used in our computations. Section 4 describes the application of persistent homology to the evaluation of the voxel carving process. We draw some conclusions and ideas for future work in the last Section.

2 Voxel Carving Approach

Space carving is a well-known method for constructing three-dimensional models of objects from a set of images. The process involves capturing a series of images of an object, and, by analysis of these images, deriving a description of the shape of the object. In particular, space (or voxel) carving aproaches [2,3,9,11] are non-invasive techniques that allow the reconstruction of a 3D human model from the images captured from a set of cameras placed around the subject. In each image, firstly, the region of interest (subject silhouette) is segmented from the background by an autonomous adaptive "approximate median" background modelling algorithm; then a 3D bounding box is drawn around the subject's approximate position in 3D space. By using extracted silhouettes from each image, inconsistent voxels are eliminated from the defined volume, iterating through each of the cameras [9]. In [1], the authors present a technique to quantitatively evaluate spatially carved volumetric representations of humans using a synthetic dataset of typical sports motion in a tennis court scenario. Such a quantification is based on the computation of Normalised Mean Square Error (NMSE) of a groundtruth volumentric reconstruction (which has been considered at 50 cameras, based on experimental observation) against any reconstruction from an inferior camera setup (with less cameras than the setup used to carve the ground truth). The aim of such an evaluation is to somehow quantify the accuracy of the 3D volume produced by the voxel carving process with regard to the number of cameras used. This investigation was motivated by the fact that very little work has been done to date on evaluating the quality of space carving results. In this paper, we intend to give a different insight into the voxel carving work by homologically characterising the sequence of reconstruction volumes. This may be interesting as the surfaces produced with a few cameras are quite noisy with many holes, which are irrelevant topological information that can be discarded by using persistent homology. Given the nature of the carvings, we believe that a homology-based approach is a more appropriate quantification than the relatively simple NMSE-based approach used previously.

3 Homology Computations on a Set of Voxels

A *cell complex* is a general topological structure by which a space is decomposed into basic elements (*cells*) of different dimensions, which are *glued* together by their boundaries (see a formal definition of CW-complex in [8]). Due to the nature of our input data, we focus on a special type of cell complex: *cubical complex*. A cubical complex Q in $\mathbf{R}^3$, is given by a finite collection of p-cubes

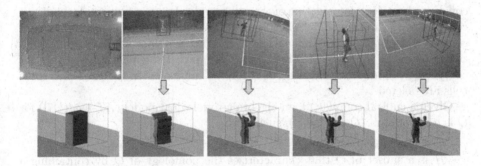

Fig. 1. Voxel carving approach for 3D reconstruction. Process with 4 cameras around the subject and an overhead camera.

such that a 0-cube is a vertex, a 1-cube is an edge, a 2-cube is a filled square (we call it, simply, a square) and a 3-cube is a filled cube (resp. a cube); together with all their faces and such that the intersection between two of them is either empty or a face of each of them.

We consider $\mathbf{Z}/2$ as the ground ring for algebraic computations, since we do not need to deal with torsion. The *cubical chain complex* associated to the cubical complex Q is the collection $\mathcal{C}(Q) = \{C_p(Q), \partial_p\}_p$ where:

(a) each $C_p(Q)$ is the corresponding chain group generated by the p-cubes of Q, over $\mathbf{Z}/2$;
(b) the boundary operator $\partial_p : C_p(Q) \rightarrow C_{p-1}(Q)$ connects two immediate dimensions. The boundary of a p-cube is the formal sum (mod 2) of all its facets (proper faces of maximal dimension). It is extended to p-chains by linearity.

Roughly speaking, the homology groups of a cubical chain complex will be a chain group whose elements are equivalence classes of *cycles*, such that if one cycle can be obtained from another by continuous deformation through the object, then they are *homologous* (or equivalent). For example, two vertices are homologous if there exists a path through the object between them. Formally, a p-cycle is a p-chain a such that $\partial_p(a) = 0$. If $a = \partial_{p+1}b$ for some $p+1$-chain b then a is called a p-boundary. We say that two p-cycles a and b are *homologous* if there exists a $(p+1)$-chain c such that $a = b + \partial_{p+1}c$. Define the p-th *homology group* to be the quotient group of p-cycles mod p-boundaries denoted by $H_p(Q)$. Each element $[a]$ of $H_p(Q)$ is a quotient class obtained by adding each p-boundary to a given p-cycle a called a *representative cycle* of the homology class $[a]$. The *homology* of Q is the chain group $\mathcal{H}(Q) = \{H_p(Q)\}_p$. See [10] for further details.

3.1   Incremental-Decremental Algorithms for Computing Persistent Homology

We focus on homology computation methods based on the concept of *AT-model* [7]. Given a cell complex, Incremental Algorithm for computing AT-models [7] computes homology information of the cell complex by an incremental technique, considering the addition of a cell each time. Once homology of an object has been computed, the same algorithm can be used again to update homology information if new cells are added to the existing complex; Decremental Algorithm for computing AT-models [6] can be used for the same aim, in the case that some cells are deleted.

Given a cubical complex Q, an algebraic-topological model (AT-$model$ [7]) for Q is a set of data (Q, H, f, g, ϕ), such that:

- Q is the cubical complex itself.
- H is a subset of Q that characterizes the homology of Q by containing a p-cube for each p-homology class, for all p. In 3D, H can only have points, edges and squares: each point of H represents a connected component of Q, each edge represents a "tunnel" and each square represents a "void" (i.e. a connected component of the background inaccessible from the outside).
- f is a chain map from $\mathcal{C}(Q)$ to $\mathcal{C}(H)$. This map provides the equivalence relation between cycles (that is, if two cycles, a and b, are equivalent, then $f(a) = f(b)$). Moreover, $f\,g(c) = c$ for any $c \in H$.
- g is a chain map from $\mathcal{C}(H)$ to $\mathcal{C}(Q)$. For each cube c in H, $g(c)$ is a representative cycle of a homology class.
- ϕ is a map from $\mathcal{C}(Q)$ to $\mathcal{C}(Q)$ that is a chain homotopy (see [10]) from $g\,f$ to the identity homomorphism on $\mathcal{C}(Q)$. This map can be seen as a kind of boundary inverse. For example, if c is a vertex, then $\phi(c)$ is the path from c to the vertex $v \in H$ homologous to c.

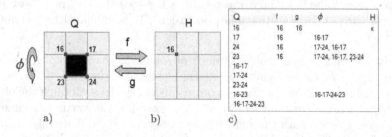

Fig. 2. A simple example of execution of Incremental Algorithm for computing AT-models. a) The input cubical complex, a filled square with all its faces (only the labels of the vertices are shown). b) The elements in H. c) The table with the information related to f, g and ϕ. Read, for instance, $f(16) = 16$, $g(16) = 16$, $\phi(16) = 0$, $\phi(17) = 16 - 17$ (edge from 16 to 17).

In [5], the authors revisit the algorithm for computing AT-models using an incremental technique that appears in [7] (we will refer to it as the Incremental Algorithm) with the aim of setting its equivalence with persistent homology computation algorithm [4,12]. Given a cubical complex Q associated to a 3D digital image, consider a full ordering of its cubes $\{c^1, \dots, c^n\}$ such that if c^i is a face of c^j, then $i < j$; take a nested sequence of subcomplexes $\emptyset = Q^0 \subseteq Q^1 \cdots \subseteq Q^n$ (a filtration over Q) such that $Q^i = \{c^1, \ldots, c^i\}$ (notice that all the proper faces of c^i are in Q^{i-1}). Under these conditions, Incremental Algorithm can be applied to compute persistent homology over the filtration.

See Fig. 2 as a simple example of execution of the Incremental Algorithm for computing AT-models.

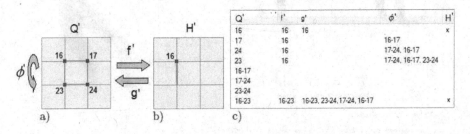

Q'	f'	g'	φ'	H'
16	16	16		x
17	16		16-17	
24	16		17-24, 16-17	
23	16		17-24, 16-17, 23-24	
16-17				
17-24				
23-24				
16-23	16-23	16-23, 23-24, 17-24, 16-17		x

a)   b)   c)

Fig. 3. A simple example of execution of Decremental Algorithm for computing an AT-model (Q', H', f', g', ϕ') after removing a 2-cube (the square $16 - 17 - 24 - 23$) from Fig. 2.a. a) The output cubical complex (only the labels of the vertices are shown) after deleting the square. b) The cubes in H'. c) The table with the information related to f', g' and ϕ'.

Now, let (Q, H, f, g, ϕ) be an AT-model for a cubical complex Q computed by the Incremental Algorithm. Let c^m be a maximal cube of Q. Then an AT-model for $Q' = Q \setminus \{c^m\}$, (Q', H', f', g', ϕ'), can be constructed by the Decremental Algorithm given in [5], where it was redefined (with respect to the one of [6]) with the aim of extending the concept of persistent homology for objects with a filtration that is not necessarily increasing.

See Fig. 3 as an example of execution of Decremental Algorithm for computing AT-models. Notice that by removing the 2-cube from the initial cubical complex on Fig. 2, a new homology class is created. The output of the algorithm is the set (Q', H', f', g', ϕ') represented in a table form in Fig. 3.c).

Now, let $\emptyset = Q^0 \leftrightarrow Q^1 \leftrightarrow \cdots \leftrightarrow Q^n$ be a *zig-zag filtration*, that is, a sequence of cell complexes such that every two consecutive complexes differ by a single cell c, i.e. either $Q^i = Q^{i-1} \cup \{c\}$ or $Q^i = Q^{i-1} \setminus \{c\}$. Then, one can compute persistent homology over the filtration by combining the application of Incremental and Decremental Algorithms depending on whether a cell c is added or deleted each time.

4 Persistent Homology for 3D Reconstruction Evaluation

We are concerned with the application of persistent homology computation to provide topological evaluation of the 3D reconstruction process by the voxel carving technique. The new insight could significantly enrich the evaluation made

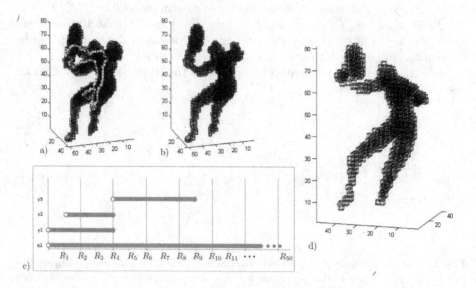

Fig. 4. 3D Reconstructions using a) 4 cameras and b) 10 cameras. Representative cycles of homology are highlighted in both cases. c) Barcode associated to the whole sequence of 3D reconstructions with increasing number of cameras (from 1 to 50). d) 3D reconstruction using 50 cameras, what is considered the groundtruth model.

in [1] by means of NMSE quantification. For this aim, we must consider the sequence of different 3D models, obtained by voxel carving under increasing number of cameras, as a whole object on which we have to set up a filtration over which to compute persistent homology. This way, in particular, we can get an estimation of the minimum number of cameras needed in order to obtain a topologically correct 3D model (which in general has only one connected component and no tunnels or voids).

We denote by R_k the cubical complex associated to the 3D reconstruction obatined using k cameras (which are randomly chosen). Starting from the first reconstruction R_1 (obtained by "one carving" of the initial 3D bounding box), we can use Incremental Algorithm to compute its homology. Notice that R_{k+1} may be obtained from R_k by removing some voxels (cubes, together with all their faces in the cubical complex). This fact makes this context good for making use of the Decremental Algorithm for getting homology computations through increasing number of cameras. Both, Incremental and Decremental Algorithms provide all the pairs of cells responsible for the creation/destruction of homology classes along the process, what allows to follow the evolution of these classes with respect to time, that is, the number of cameras used. Actually, to compute persistent homology of the whole sequence of 3D models, $\{R_k\}_k$, the zigzag filtration is given by the sequence $\{R_k\}_k$ itself with the inclusion, between R_k and R_{k+1}, of a sequence of complexes $\{R_k^{i_k}\}_{i_k=1...n_k}$ given by the addition or deletion of a cell, each time. Compute, then, a big barcode for visualizing the hole computation in

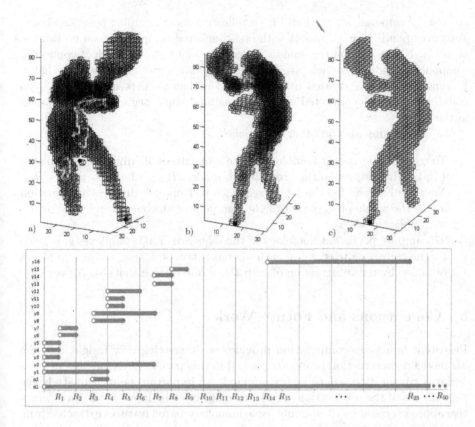

Fig. 5. 3D Reconstructions (viewed from different angles) using different number of cameras: a) 4 cameras, b) 15 cameras and c) 24 cameras, which is similar to the one obtained with 50 cameras (groundtruth model). Representative cycles of homology are highlighted. Below, barcode associated to the whole sequence of 3D reconstructions from 1 to 50 cameras.

order to easily analyze the stability of the elements of homology. We want also to remark that, due to the nature of the voxel carving process, only voxels on the surface of the object are removed each time, so different connected componentes and tunnels (but no cavities) may arise.

We have used for computation five different frames extracted from a 3D video sequence with a voxel resolution of 4 cm, that is, the spacing between each voxel is 4 cm in the OX, OY and OZ directions. This means $15,625$ voxels per cubic metre. We have appreciated, as it was expected, that simpler poses of the subject produce simpler barcodes while more complex poses give place to more interesting homological information. Fig. 4 shows that the carving process, in a case of simple pose, stabilizes at 10 cameras (with a unique connected component), while below that point, 3 different tunnels have been living for some time. That means that, in order to produce a topologically correct model, at least 10 cameras are needed. Fig. 5 reflects a more complex pose, though it also corresponds to a 3D object with one connected component and no tunnels or voids. Notice the more complex barcode associated (in which 2 connected components and 16 tunnels are represented) and, especially, the fact that a 1–homology class is created at time $k = 15$, that persists until $k = 23$. So stabilization of one connected component as final state, occurs much later than in the former case.

We are working also on other approaches:

- To compute persistent homology of the sequence of 3D difference complexes $\{D_k\}_k$ with respect to the groundtruth model (R_∞), where $D_k = R_k \setminus R_\infty$. Now the barcode for the whole sequence will provide different information about the whole process that might complement the one given by the reconstructions themselves.
- To compute persistent homology of the sequence of 3D complexes generated by the convex deficiencies of each 3D reconstruction, that is, the complexes obtained by the substraction of each 3D reconstruction to its 3D convex hull.

5   Conclusions and Future Work

Persistent homology computation provides an interesting new insight into the 3D model reconstruction process explained in this paper. There are lots of ideas and experimentation still to be investigated. An important point is to study the dependence of the observations on the resolution of the input data. An alternative approach could be to compute some homology-based features extracted from each reconstruction R_k and to compare them against a groundtruth model. These features should be measurable so that a distance with respect to the groundtruth model could be computed. These parameters could be extracted from the comparison of weighted histograms of connected components (for 0-homology study) or minimal-legth (in some sense) representative cycles of 1-homology. Another interesting question arises by fixing a certain number of cameras and considering the sequence of 3D reconstructions along time. Holes can be produced along time by different movements of the body that should not be stable along the complete scene, so the homological analysis of voxel carving performance of video sequences could shed some light on the classification of these movements.

References

1. Monaghan, D., Kelly, P., OConnor, N.E.: Quantifying Human Reconstruction Accuracy for Voxel Carving in a Sporting Environment. In: ACM MM, Scottsdale, AZ, November 28 - December 1 (2011)
2. Broadhurst, A., Drummond, T., Cipolla, R.: A probabilistic framework for space carving. In: Conf. on Computer Vision, vol. 1, p. 388 (2001)
3. Culbertson, W.B., Malzbender, T., Slabaugh, G.: Generalized voxel coloring. In: Intern. Workshop on Vision Algorithms: Theory and Practice, pp. 100–115 (1999)

4. Edelsbrunner, H., Letscher, D., Zomorodian, A.: Topological persistence and simplification. In: FOCS 2000, pp. 454–463. IEEE Computer Society (2000)
5. Gonzalez-Diaz, R., Ion, A., Jimenez, M.J., Poyatos, R.: Incremental-Decremental Algorithm for Computing AT-Models and Persistent Homology. In: Real, P., Diaz-Pernil, D., Molina-Abril, H., Berciano, A., Kropatsch, W. (eds.) CAIP 2011, Part I. LNCS, vol. 6854, pp. 286–293. Springer, Heidelberg (2011)
6. Gonzalez-Díaz, R., Medrano, B., Sánchez-Peláez, J., Real, P.: Simplicial Perturbation Techniques and Effective Homology. In: Ganzha, V.G., Mayr, E.W., Vorozhtsov, E.V. (eds.) CASC 2006. LNCS, vol. 4194, pp. 166–177. Springer, Heidelberg (2006)
7. Gonzalez-Diaz, R., Real, P.: On the cohomology of 3D digital images. Discrete Applied Math. 147(2-3), 245–263 (2005)
8. Hatcher, A.: Algebraic Topology. Cambridge University Press (2002)
9. Kutulakos, K.N., Seitz, S.M.: A theory of shape by space carving. Intern. Journal of Computer Vision 38, 199–218 (2000)
10. Munkres, J.: Elements of Algebraic Topology. Addison-Wesley Co. (1984)
11. Seitz, S.M., Curless, B., Diebel, J., Scharstein, D., Szeliski, R.: A comparison and evaluation of multi-view stereo reconstruction algorithms. In: IEEE Conference on Computer Vision and Pattern Recognition, vol. 1, pp. 519–528 (2006)
12. Zomorodian, A., Carlsson, G.: Computing persistent homology. Discrete and Computational Geometry 33(2), 249–274 (2005)

Persistence Modules, Shape Description, and Completeness

Francesca Cagliari[1], Massimo Ferri[1,3],
Luciano Gualandri[1], and Claudia Landi[2,3]

[1] Dip. di Matematica, Univ. di Bologna, Italy
{francesca.cagliari,massimo.ferri,luciano.gualandri}@unibo.it
[2] Dip. di Scienze e Metodi dell'Ingegneria, Univ. di Modena e Reggio Emilia, Italy
claudia.landi@unimore.it
[3] ARCES, Univ. di Bologna, Italy

Abstract. Persistence modules are algebraic constructs that can be used to describe the shape of an object starting from a geometric representation of it. As shape descriptors, persistence modules are not complete, that is they may not distinguish non-equivalent shapes. In this paper we show that one reason for this is that homomorphisms between persistence modules forget the geometric nature of the problem. Therefore we introduce geometric homomorphisms between persistence modules, and show that in some cases they perform better. A combinatorial structure, the H_0-tree, is shown to be an invariant for geometric isomorphism classes in the case of persistence modules obtained through the 0th persistent homology functor.

Keywords: geometric homomorphism, rank invariant, H_0-tree.

1 Introduction

The shape description problem is at the core of many shape recognition methods used in computer vision and computer graphics. It is based on the fundamental idea of using compact representations of shapes, namely, shape descriptors, to analyze, understand, and compare objects [17]. Roughly speaking, a shape descriptor is complete when any two different shapes have two different descriptions.

In this paper we consider persistence modules as shape descriptors, focusing on the completeness problem, that is the problem of deciding whether persistence modules are able to discriminate between different shapes. We highlight some differences that arise in this respect depending on whether we work in a purely algebraic setting or we also keep memory of the underlying geometric setting to define the set of homomorphisms between persistence modules.

Approaching the shape description problem by persistence, one usually models the shape of an object as a geometric pair (X, f), where X is a geometric representation of the object under study (e.g., a manifold or a triangular mesh), and f is a function on X measuring some shape property of the object (e.g., curvature, height, distance from a fixed point). Two objects are considered to have the same shape whenever the pairs (X, f) and (X', f') modeling

M. Ferri et al. (Eds.): CTIC 2012, LNCS 7309, pp. 148–156, 2012.

them are isomorphic in a suitable category $\mathcal{C}$ [12]. However, since answering the isomorphism question in $\mathcal{C}$ is not an easy problem, one is usually satisfied with moving to persistence modules via the persistent homology functor, and studying the isomorphism question for persistence modules [18,4]. In this way the shape descriptors that are actually used for comparison are persistence modules.

The rationale behind this approach is that the persistent homology functor does not change the isomorphism classes. Moreover, the isomorphism question for persistence modules is easier than for objects in $\mathcal{C}$. Sometimes one simplifies further the problem by only considering as shape descriptors invariants of isomorphism classes of persistence modules, such as barcodes [5], or size functions [7].

The completeness problem for persistence modules is studied in [11] for the case of curves. The authors show that two different shapes can have isomorphic persistence modules (i.e. the persistent homology functor forgets some relevant geometric features of the original shape), and prove that completeness can be achieved by increasing the number of components of the measuring function f.

In this paper, after reporting the basic definitions about persistence in Sect. 2, in Sect. 3 we present an example showing that, in some cases, persistence modules cannot discriminate non-equivalent shapes because in the category of persistence modules there are homomorphisms that are purely algebraic. More precisely, they do not reflect geometric transformations between the original shapes. For this reason, in Sect. 4, we introduce the notion of geometric homomorphisms between persistence modules and show that in some case they perform better than algebraic homomorphisms. Finally, in Sect. 5, we consider the problem of completeness for invariants. As for invariants with respect to algebraic isomorphisms, we review some results about the rank invariant from [2]. In the case of invariants with respect to geometric isomorphisms, we present an invariant in the case of 0th homology, the H_0-tree. We end the paper with a brief list of open questions in Sect. 6.

2   Background on Persistence

2.1   The Geometric Approach to Persistence

According to the spirit of the original persistence papers [10,1], one has some category $\mathcal{C}$ of interest of geometric nature, and a functor F from that category to the category n-filt of n-filtrations. One then studies and works with the functor $H_i \circ F$, H_i being the ordinary ith homology functor. This composite functor is generally called an *ith persistent homology functor*.

In the most simple case, fixed $n \in \mathbb{N}$, $\mathcal{C} = \mathcal{C}(n)$ is the category defined as follows [15]:

1. Objects of $\mathcal{C}$ are pairs (X, f), where X is a topological space and $f = (f_1, \ldots, f_n) \colon X \to \mathbb{R}^n$ is a continuous function.

2. If $(X, f), (X', f') \in obj(\mathcal{C})$, then the set of morphisms of $\mathcal{C}$ from (X, f) to (X', f') is the set of continuous functions $\gamma \colon X \to X'$ such that $f(x) \geq f'(\gamma(x))$ for all $x \in X$ (with the convention that $u = (u_i) \leq v = (v_i)$ in $\mathbb{R}^n$ means $u_i \leq v_i$ for all i).

We observe that an isomorphism in $\mathcal{C}$ is a homeomorphism mapping level sets into level sets. Therefore we obtain the following easy remark.

Proposition 1. *If two pairs (X, f) and (X', f') are isomorphic in $\mathcal{C}$ then their natural pseudo-distance vanishes:*

$$\delta((X, f), (X', f')) := \inf_{\gamma} \sup_{x \in X} \|f(x) - f(\gamma(x))\|_\infty = 0,$$

where γ varies in all possible homeomorphisms between X and X'.

More details on the natural pseudo-distance and its relationship with persistence can be found, e.g., in [6,8]. We remark that the converse of Proposition 1 in general is false, although it can be true in certain cases (cf., e.g., [3]).

For $(X, f) \in obj(\mathcal{C})$, and for $u = (u_i) \in \mathbb{R}^n$, let $X_u = \cap_{i=1}^{n} f_i^{-1}((-\infty, u_i])$. If $u \leq v \in \mathbb{R}^n$, then there is an inclusion $i_X(u, v) \colon X_u \hookrightarrow X_v$. Thus the collection $\{X_r\}_{r \in \mathbb{R}^n}$ is an n-filtration of X. If $(X, f), (X', f') \in obj(\mathcal{C})$, and $\gamma \colon X \to X'$ is a morphism from (X, f) to (X', f'), then the restriction of γ to X_u, denoted by γ_u, maps X_u to X'_u, for all $u \in \mathbb{R}^n$. Moreover, for all $u \leq v \in \mathbb{R}^n$, $\gamma_v \circ i_X(u, v) = i_{X'}(u, v) \circ \gamma_u$. Thus the collection $\{\gamma_r\}_{r \in \mathbb{R}^n}$ is a morphism of n-filtrations. The functor $F \colon \mathcal{C} \to n$-filt which maps (X, f) to $\{X_r\}_{r \in \mathbb{R}^n}$ and γ to $\{\gamma_r\}_{r \in \mathbb{R}^n}$ is called the *sublevelset filtration functor*.

2.2   The Algebraic Approach to Persistence

In [2,18], the authors showed that persistence can be defined at algebraic level directly, without the need for an underlying topological setting. More precisely, they introduced the concept of a *persistence module* $\mathbf{M}$ as the one of a family $\{M_u\}_{u \in \mathbb{R}^n}$ of vector spaces (or modules over the same commutative ring) together with a family of homomorphisms $\{\iota_M(u, v) \colon M_u \to M_v\}_{u \leq v \in \mathbb{R}^n}$ such that $\iota_M(u, w) = \iota_M(v, w) \circ \iota_M(u, v)$ and $\iota_M(u, u) = \mathrm{id}_{M_u}$ for all $u \leq v \leq w \in \mathbb{R}^n$.

Given two persistence modules $\mathbf{M}$ and $\mathbf{N}$, the set of homomorphisms from $\mathbf{M}$ to $\mathbf{N}$ consists of collections of homomorphisms of vector spaces $\mathbf{h} = \{h_u \colon M_u \to N_u\}_{u \in \mathbb{R}^n}$ such that $\iota_N(u, v) \circ h_u = h_v \circ \iota_M(u, v)$ for all $u \leq v \in \mathbb{R}^n$. Therefore, in a purely algebraic setting two persistence modules $\mathbf{M}$ and $\mathbf{N}$ are isomorphic if there exists a collection $\mathbf{h} = \{h_u \colon M_u \to N_u\}_{u \in \mathbb{R}^n}$ of isomorphisms of vector spaces such that $\iota_N(u, v) \circ h_u = h_v \circ \iota_M(u, v)$ for all $u \leq v \in \mathbb{R}^n$. We call $\mathbf{h}$ an *algebraic isomorphism of persistence modules*.

The category of persistence modules will be denoted by $\mathcal{M}$. Clearly, objects and homomorphisms of $\mathcal{M}$ can be constructed by applying the persistent homology functor to objects and morphisms of $\mathcal{C}$. It is known that, in the case of objects, the converse is also true, at least for finite persistence modules [2, Th. 2]. For this reason, in this paper we focus on morphisms rather than on objects of $\mathcal{M}$.

3   A Preliminary Example

Let us begin considering the following example. It shows that, even in the very basic case of curves endowed with simple Morse functions, we can find non-isomorphic pairs (X, f) and (X', f') in $\mathcal{C}$ taken by the persistent homology functor into algebraically isomorphic modules.

Example 1. Let (X, f) and (X', f'), with $X = X' = S^1$, be the two curves displayed in Figure 1. (X, f) and (X', f') are not isomorphic in $\mathcal{C}$. Indeed, an isomorphism between (X, f) and (X', f') necessarily takes a critical point of f to the critical point of f' at the same height, which is clearly impossible in this case. On the other hand, the persistence modules $\mathbf{M}$ and $\mathbf{N}$ obtained by applying the 0th persistent homology functor to the pairs (X, f) and (X', f'), respectively, are isomorphic in $\mathcal{M}$. To see this, it is sufficient to consider the diagram

$$
\begin{array}{ccc}
M_5 = < z_1, z_2, z_3 | z_1 = z_2 = z_3 > & \xrightarrow{h_5} & N_5 = < z_1', z_2', z_3' | z_1' = z_2' = z_3' > \\
\uparrow & & \uparrow \\
M_4 = < z_1, z_2, z_3 | z_1 = z_3 > & \xrightarrow{h_4} & N_4 = < z_1', z_2', z_3' | z_2' = z_3' > \\
\uparrow & & \uparrow \\
M_3 = < z_1, z_2, z_3 > & \xrightarrow{h_3} & N_3 = < z_1', z_2', z_3' > \\
\uparrow & & \uparrow \\
M_2 = < z_1, z_2 > & \xrightarrow{h_2} & N_2 = < z_1', z_2' > \\
\uparrow & & \uparrow \\
M_1 = < z_1 > & \xrightarrow{h_1} & N_1 = < z_1' >,
\end{array}
\tag{1}
$$

where the vertical maps are induced by inclusions, and the horizontal maps are defined by setting $h_1(z_1) = z_1'$, $h_2(z_1) = z_1'$, $h_2(z_2) = z_2'$, $h_3(z_1) = z_1'$, $h_3(z_2) = z_2'$, $h_3(z_3) = z_1' + z_2' - z_3'$, $h_4(z_1) = z_1'$, $h_4(z_2) = z_2'$, $h_5(z_1) = z_1'$.

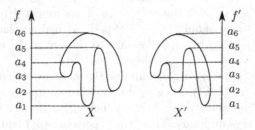

Fig. 1. Two curves not distinguishable by persistence modules in $\mathcal{M}$

The previous example proves the following result, saying that algebraic isomorphism of persistence modules may not distinguish the shapes of two pairs.

Proposition 2. *The persistent homology functor $H_i \circ F : \mathcal{C} \to \mathcal{M}$ does not reflect isomorphisms (i.e., $H_i \circ F(X, f)$ isomorphic to $H_i \circ F(X', f')$ in $\mathcal{M}$ does not imply (X, f) isomorphic to (X', f') in $\mathcal{C}$).*

This prompts a new definition that will be given in the next section.

4   Geometric Homomorphisms

In this section we focus on particular homomorphisms between persistence modules that we call *geometric* because they are the image of a morphism in $\mathcal{C}$. In Proposition 3 we prove that, restricting to geometric homomorphisms, we can distinguish the curves of Example 1.

Definition 1. *Let (X, f) and (X', f') be two objects in $\mathcal{C}$. A homomorphism (resp., isomorphism) $\mathbf{h}$ between the persistence modules $H_i \circ F(X, f)$ and $H_i \circ F(X', f')$ is called a* geometric homomorphism *(resp. geometric isomorphism) if it belongs to the image of the persistent homology functor.*

We now prove that no isomorphism between the persistence modules $\bigoplus_{i \in \mathbb{Z}} H_i \circ F(X, f)$ and $\bigoplus_{i \in \mathbb{Z}} H_i \circ F(X', f')$ of Example 1 belongs to the image of the persistent homology functor.

Proposition 3. *Let (X, f) and (X', f') be as in Example 1. No morphism γ between (X, f) and (X', f') is taken by $\bigoplus_{i \in \mathbb{Z}} H_i \circ F$ into an isomorphism of persistence modules.*

Proof. Let $\gamma : X \to X'$ be a continuous function such that $f(x) \geq f'(\gamma(x))$ for every $x \in X$ (i.e., γ is a morphism in $\mathcal{C}$). Let us assume that $H_i \circ F(\gamma)$ is an isomorphism for every $i \in \mathbb{Z}$. Hence, in particular, $H_1(\gamma) : H_1(X) \to H_1(X')$ is an isomorphism, and therefore the degree of γ is non-zero (recall that $X = X' = S^1$). It follows that $\gamma(X) = X'$.

Let a_i with $i = 1, \ldots, 6$ be as in Figure 1 and let p_i (resp. q_i) be the only critical point of f in $f^{-1}(a_i)$ (resp. of f' in $(f')^{-1}(a_i)$). Since $f(p_1) \geq f'(\gamma(p_1))$, necessarily $\gamma(p_1) = q_1$. Moreover, it must hold $\gamma(p_6) = q_6$. Indeed, since $\gamma(X) = X'$, there is some $p \in X$ such that $\gamma(p) = q_6$. Thus we get $f(p) \geq f(\gamma(p)) = a_6$, implying $p = p_6$. Using again $f(x) \geq f'(\gamma(x))$, we deduce that $f'(\gamma(p_2)) \leq a_2$. Hence, either $\gamma(p_2) = q_2$ or $\gamma(p_2)$ belongs to the arc of X' containing q_1 and staying under a_2.

Let us assume $\gamma(p_2) = q_2$. By considering the arcs in which p_1, p_6 (resp. q_1, q_6) split the curve X (resp. X'), since p_3 does not belong to the arc containing p_2, by continuity we get that $\gamma(p_3)$ does not belong to the arc containing $\gamma(p_2)$. Moreover, $\gamma(p_3)$ stays under a_3. Thus, the classes of $\gamma(p_1)$ and $\gamma(p_3)$ are homologous in $N_3 = H_0(X'_{a_3})$. Since the classes of p_1 and p_3 are not homologous in $M_3 = H_0(X_{a_3})$, we conclude that $H_0 \circ F(\gamma)$ is not an isomorphism.

Otherwise, if $\gamma(p_2)$ belongs to the arc of X' containing q_1 and staying under a_2, then $\gamma(p_1)$ and $\gamma(p_2)$ are homologous in $N_2 = H_0(X'_{a_2})$. Since the classes of p_1 and p_2 are not homologous in $M_2 = H_0(X_{a_2})$, we conclude that $H_0 \circ F(\gamma)$ is not an isomorphism in this case either, yielding the claim.   □

Thus, if we consider the subset of geometric isomorphisms we can distinguish the curves of Example 1. This seems to suggest that the image of the persistent homology functor is better suited for the aims of shape comparison than persistence modules.

5   Invariants

In this section we study invariants for isomorphism classes of persistence modules in the algebraic as well as in the geometric case.

5.1   Algebraic Setting

Invariants for classes of persistence modules up to algebraic isomorphism have been thoroughly studied for cases $n = 1$ in [18] and $n > 1$ in [2]. The main invariant proposed is the *rank invariant*.

Definition 2. *Given a persistence module* $\mathbf{M}$ *consisting of vector spaces* $\{M_u\}_{u \in \mathbb{R}^n}$ *and homomorphisms* $\{\iota_M(u,v) : M_u \to M_v\}_{u \leq v \in \mathbb{R}^n}$, *its rank invariant is an integer-valued function* $\rho_{\mathbf{M}}$ *of two variables* $u \leq v \in \mathbb{R}^n$, *defined by* $\rho_{\mathbf{M}}(u,v) = \mathrm{rk}\,(\iota_M(u,v))$.

In [18] the authors show that, for $n = 1$, the rank invariant is a complete invariant for algebraic isomorphism of persistence modules admitting a finite presentation (in terms of generators and relators). This means that any two such persistence modules are algebraically isomorphic if and only if their rank invariants coincide.

The analogous property for $n > 1$ is false, as the following example shows (see also [2]).

Example 2. Consider the bi-dimensional persistence modules $\mathbf{M}$ and $\mathbf{N}$, given by

$$M_{1,3} = < z_b >$$
$$M_{2,2} = < z_a >$$
$$M_{3,1} = < z_a >,$$
$$\begin{pmatrix} 0 & 1 \end{pmatrix}$$
$$\begin{pmatrix} 1 & 1 \end{pmatrix}$$
$$M_{0,0} = < z_a, z_b > \underline{\qquad\qquad} \begin{pmatrix} 1 & 0 \end{pmatrix} \underline{\qquad\qquad}$$

$$N_{1,3} = < z'_a >$$
$$N_{2,2} = < z'_b >$$
$$N_{3,1} = < z'_a >.$$
$$\begin{pmatrix} 1 & 0 \end{pmatrix}$$
$$\begin{pmatrix} 0 & 1 \end{pmatrix}$$
$$N_{0,0} = < z'_a, z'_b > \underline{\qquad\qquad} \begin{pmatrix} 1 & 1 \end{pmatrix} \underline{\qquad\qquad}$$

where the row matrix displayed on each arrow represents the homomorphism between the modules connected by the arrow, with respect to the bases enclosed by angle brackets.

For instance, $\mathbf{M}$ and $\mathbf{N}$ can be obtained by applying the 1st homology functor with coefficients in $\mathbb{Z}_2$ to the 2-filtrations displayed in Fig. 2.

$\mathbf{M}$ and $\mathbf{N}$, as persistence modules over $\mathbb{Z}_2$, are not isomorphic, although their rank invariants coincide.

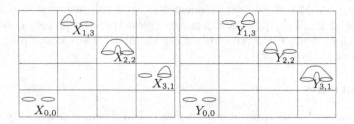

Fig. 2. Filtrations taken by the 1st homology functor into **M** (left) and **N** (right)

5.2 Geometric Setting

We now consider invariants for classes of persistence modules up to geometric isomorphism. We confine our treatment to the case $n = 1$, that is to scalar functions $f: X \to \mathbb{R}$. In this setting we can consider the so-called H_0-tree of f introduced in [1].

Trees are of widespread use in topology, either for invariant computations in a discrete setting (e.g., spanning trees for the fundamental group [13], and spanning forests for homology [16]) or as signatures in a continuous setting (e.g., contour trees [14] for domains of the plane, and merge trees [9] for arbitrary manifolds). The use we make of trees in this section is in the latter spirit.

H_0-trees can be defined on any topological space endowed with a (suitable) scalar function used to filter the space. Intuitively, the connected components of the sub-level sets of a function f, thanks to the inclusion relation, can be organized in a directed tree structure where node parenthood maps component inclusion. We shall prove that H_0-trees are invariant for geometric isomorphism classes of 0th homology.

Definition 3. *For a closed (i.e. compact and without boundary) connected manifold X and a simple Morse function $f: X \to \mathbb{R}$, the H_0-tree of f is a rooted binary tree labeled on the nodes defined as follows:*

- *the set of nodes is equal to the set of points of X such that for every sufficiently small real value $\varepsilon > 0$, the homomorphism induced by the inclusion $\iota(f(p) - \varepsilon, f(p) + \varepsilon): H_0(X_{f(p)-\varepsilon}) \to H_0(X_{f(p)+\varepsilon})$ is not an isomorphism;*
- *the label of a node p is equal to $f(p)$;*
- *p is a child of q if q has the lowest label among the nodes for which $f(p) < f(q)$ and $\iota(f(p), f(q)): H_0(X_{f(p)}) \to H_0(X_{f(q)})$ takes the class of p to that of q.*

The H_0-trees corresponding to the curves of Fig. 1 are displayed in Fig. 3.

Proposition 4. *Let X, X' be closed connected manifolds, and $f: X \to \mathbb{R}$, $f': X' \to \mathbb{R}$ be simple Morse functions. If $\mathbf{h}$ is a geometric isomorphism between $\mathbf{M} = H_0 \circ F(X, f)$ and $\mathbf{N} = H_0 \circ F(X', f')$, then the H_0-trees of f and f', say T and T', are isomorphic as labeled trees.*

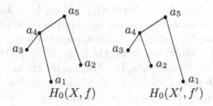

$$H_0(X, f) \qquad H_0(X', f')$$

Fig. 3. The H_0-trees associated with the two curves of Fig. 1

Proof. Since **h** is an isomorphism, there is a label preserving bijection σ between the set of nodes of T and that of T'. Let us see that σ also preserves the edges. Let p be a child of q in T. Then $\iota(f(p), f(q))$ sends the class of p into that of q. Since **h** is induced by a morphism $\gamma: X \to X'$, $\iota(f'(\gamma(p)), f'(\gamma(q)))$ sends the class of $\gamma(p)$ into that of $\gamma(q)$, and the class of $\gamma(p)$ (resp. $\gamma(q)$) coincides with that of $\sigma(p)$ (resp. $\sigma(q)$). Therefore $\sigma(p)$ is a child of $\sigma(q)$. $\qquad \square$

We remark that H_0-trees are not invariant for algebraic isomorphism classes of 0th persistence modules. Indeed, the curves of Example 1 have non-isomorphic H_0-trees whereas their persistence modules are algebraically isomorphic.

We can see that H_0-trees are not complete invariants for geometric isomorphism classes by taking the examples in Fig. 1 with $-f$ and $-f'$ instead of f and f'. However, in the case $X = S^1$, we can prove that a curve (X, f), $f: X \to \mathbb{R}$ being a simple Morse function, can be completely reconstructed up to f-preserving homeomorphisms, from the H_0-trees of f and $-f$.

6 Open Questions

We think that further investigations on the subject presented here could tackle the following problems:

1. Is it true that the image of the persistent homology functor is a category?
2. Does a simple characterization of geometric homomorphisms exists?
3. Can H_0-trees be generalized so to obtain combinatorial structures providing invariants for geometric isomorphism classes for other homology degrees?

References

1. Cagliari, F., Ferri, M., Pozzi, P.: Size functions from the categorical viewpoint. Acta Appl. Math. 67, 225–235 (2001)
2. Carlsson, G., Zomorodian, A.: The theory of multidimensional persistence. Discrete Comput. Geom. 42(1), 71–93 (2009)
3. Cerri, A., Di Fabio, B.: On certain optimal diffeomorphisms between closed curves. Forum Mathematicum (accepted for publication)
4. Chazal, F., Cohen-Steiner, D., Glisse, M., Guibas, L.J., Oudot, S.Y.: Proximity of persistence modules and their diagrams. In: Proc. SCG 2009, pp. 237–246 (2009)

5. Collins, A., Zomorodian, A., Carlsson, G., Guibas, L.: A barcode shape descriptor for curve point cloud data. Computers and Graphics 28, 881–894 (2004)
6. d'Amico, M., Frosini, P., Landi, C.: Natural pseudo-distance and optimal matching between reduced size functions. Acta Appl. Math. 109, 527–554 (2010)
7. Dibos, F., Frosini, P., Pasquignon, D.: The use of size functions for comparison of shapes through differential invariants. J. Math. Imag. Vis. 21, 107–118 (2004)
8. Donatini, P., Frosini, P.: Natural pseudodistances between closed manifolds. Forum Math. 16(5), 695–715 (2004)
9. Edelsbrunner, H., Harer, J.: Computational Topology: An Introduction. American Mathematical Society (2009)
10. Edelsbrunner, H., Letscher, D., Zomorodian, A.: Topological persistence and simplification. Discrete Comput. Geom. 28(4), 511–533 (2002)
11. Frosini, P., Landi, C.: Uniqueness of models in persistent homology: the case of curves. Inverse Problems 27, 124005 (2011)
12. Frosini, P., Landi, C.: Size theory as a topological tool for computer vision. Patt. Recog. Image Anal. 9, 596–603 (1999)
13. Hatcher, A.: Algebraic topology. Cambridge University Press (2002)
14. van Kreveld, M., van Oostrum, R., Bajaj, C., Pascucci, V., Schikore, D.: Contour trees and small seed sets for isosurface traversal. In: Proc. 13th Annu. ACM Sympos. Comput. Geom., pp. 212–220 (1997)
15. Lesnick, M.: The optimality of the interleaving distance on multidimensional persistence modules (2011), http://arxiv.org/abs/1106.5305
16. Molina-Abril, H., Real, P.: Homological Computation Using Spanning Trees. In: Bayro-Corrochano, E., Eklundh, J.-O. (eds.) CIARP 2009. LNCS, vol. 5856, pp. 272–278. Springer, Heidelberg (2009)
17. Veltkamp, R.C., Hagendoorn, M.: State-of-the-art in shape matching. In: Principles of Visual Information Retrieval, pp. 87–119. Springer (2001)
18. Zomorodian, A., Carlsson, G.: Computing persistent homology. Discrete Comput. Geom. 33(2), 249–274 (2005)

Author Index

Ayala, Rafael   11

Berciano, Ainhoa   39
Brendel, Piotr   117

Cagliari, Francesca   148
Carnero, Javier   108
Cerri, Andrea   128
Čomić, Lidija   98

Damiand, Guillaume   20
De Floriani, Leila   79, 98
Dénès, Maxime   49
Díaz-Pernil, Daniel   39
Di Fabio, Barbara   128
Dłotko, Paweł   68, 117

Escribano, Carmen   88

Fernández-Ternero, Desamparados   11
Ferri, Massimo   148

Giraldo, Antonio   88
Gonzalez-Diaz, Rocio   20
Gualandri, Luciano   148
Gutierrez, Antonio   139
Gutiérrez-Naranjo, Miguel A.   39

Heras, Jónathan   49

Jiménez, María José   139

Landi, Claudia   148

Magillo, Paola   79
Mata, Gadea   49
Mazo, Loïc   1
Medri, Filippo   128
Mesmoudi, Mohammed Mostefa   79
Molina-Abril, Helena   108
Monaghan, David   139
Mörtberg, Anders   49
Mrozek, Marian   68, 117

O'Connor, Noel E.   139

Pacheco, Ana   58
Peltier, Samuel   20
Peña-Cantillana, Francisco   39
Poza, María   49

Real, Pedro   58, 108

Sastre, María Asunción   88
Siles, Vincent   49

Vilches, José Antonio   11
Vuçini, Erald   30

Wagner, Hubert   68

Żelazna, Natalia   117